The Politics of 21st Century Environmental Disasters

This book explores the main concerns for grappling with increasing environmental disasters and examines how environmental disasters are understood by states, corporations, and non-government organizations nationally and internationally.

The focus of this book is threefold: first, to investigate what constitutes an environmental disaster and to identify the parameters for political responses nationally and internationally. Second, the chapters analyse contemporary state practices that exacerbate the impact of, and responses to, environmental disasters. They show how states promote extractivism based on limited understandings of nature drawn from Western philosophy. Finally, the book highlights the strengths and weaknesses in political and institutional responses at the local level to such disasters by state and non-state actors. This shows how both slow and fast violence of environmental disasters affects communities, but also how vulnerable subjects are based on people's capabilities.

The Politics of 21st Century Environmental Disasters is an indispensable resource for students and scholars in political science and environmental studies. The chapters in this book were originally published as a special issue of *Environmental Politics*.

Susan Park is Professor of Global Governance in the Discipline of Government and International Relations at the University of Sydney. She focuses on how international organizations and global governance can become greener and more accountable, particularly in the transition to renewable energy.

The Politics of 21st Century Environmental Disasters

Edited by
Susan Park

Routledge
Taylor & Francis Group
LONDON AND NEW YORK

First published 2023
by Routledge
4 Park Square, Milton Park, Abingdon, Oxon, OX14 4RN

and by Routledge
605 Third Avenue, New York, NY 10158

Routledge is an imprint of the Taylor & Francis Group, an informa business

British Library Cataloguing in Publication Data
A catalogue record for this book is available from the British Library

ISBN13: 978-1-032-49674-0 (hbk)
ISBN13: 978-1-032-49675-7 (pbk)
ISBN13: 978-1-003-39495-2 (ebk)

DOI: 10.4324/9781003394952

Typeset in Minion Pro
by Newgen Publishing UK

Publisher's Note
The publisher accepts responsibility for any inconsistencies that may have arisen during the conversion of this book from journal articles to book chapters, namely the inclusion of journal terminology.

Contents

Citation Information

The chapters in this book were originally published in the journal *Environmental Politics*, volume 31, issue 1 (2022). When citing this material, please use the original page numbering for each article, as follows:

Introduction

The politics of 21st century environmental disasters
Susan Park
Environmental Politics, volume 31, issue 1 (2022), pp. 1–7

Chapter 1

The role of the Sovereign state in 21st century environmental disasters
Susan Park
Environmental Politics, volume 31, issue 1 (2022), pp. 8–27

Chapter 2

Planetary disasters: moving the UN disaster risk reduction framework into cosmopolitised reality
Sabine Selchow
Environmental Politics, volume 31, issue 1 (2022), pp. 28–48

Chapter 3

Rehabilitating Ranger uranium mine: scientific uncertainty, deep futures and the production of ignorance
Rebecca Lawrence
Environmental Politics, volume 31, issue 1 (2022), pp. 49–69

Chapter 4

Disaster? No surprise
Christine J. Winter
Environmental Politics, volume 31, issue 1 (2022), pp. 70–88

Chapter 5

Extractive industry disasters and community responses: a typology of vulnerable subjects
Teresa Kramarz
Environmental Politics, volume 31, issue 1 (2022), pp. 89–109

For any permission-related enquiries please visit:
www.tandfonline.com/page/help/permissions

Notes on Contributors

Teresa Kramarz, Munk School of Global Affairs and Public Policy, University of Toronto, Canada.

Rebecca Lawrence, Sydney Environment Institute, University of Sydney, Australia; Department of Political Science, Stockholm University, Sweden; Stockholm Environment Institute, Sweden.

Susan Park, Department of Government and International Relations, University of Sydney, Australia.

Sabine Selchow, School of School of Philosophical and Historical Inquiry, University of Sydney, Australia.

Christine J. Winter, Department of Government and International Relations, University of Sydney, Australia.

INTRODUCTION

The politics of 21st century environmental disasters

Susan Park

ABSTRACT

We outline the main concerns for grappling with increasing environmental disasters in this Symposium. Our focus is threefold: first, to theoretically investigate what constitutes an environmental disaster and identify the parameters for political responses through the discourse in high level multilateral fora; through the construction of the sovereign state system; and through the promotion of ignorance. Second, we aim to identify contemporary practices of the state that exacerbate the impact of, and responses to, environmental disasters. We show how states promote extractivism based on limited understandings of nature drawn from Western political liberal philosophy. Finally, we highlight the strengths and weaknesses in political and institutional responses at the local level to such disasters by state and non-state actors. This shows how both slow and fast violence of environmental disasters affects communities, but also how vulnerable subjects are predicated on pre-existing capabilities.

Introduction

We are at a critical juncture in human history. The latest Intergovernmental Panel on Climate Change (IPCC) assessment report states that it is 'unequivocal that human influence has warmed the atmosphere, ocean, and land,' and that the 'scale of recent changes across the climate system are unprecedented over many centuries to many thousands of years' (2021, p. 9). We know this has led to an increase in average surface temperature, sea level rise, and melting glaciers. It has also directly contributed to 'many weather and climate extremes in every region of the globe.' Scientific knowledge for attributing change to human activity has strengthened in relation to 'observed changes in extremes such as heatwaves, heavy precipitation, droughts, and tropical cyclones.' Human influence 'has likely increased the chance of extreme compound events since the 1950s' such as 'concurrent heatwaves and droughts, fire weather *in some regions of all inhabited continents*, and compound flooding in some locations' (IPCC 2021, pp. 10–11; emphasis added). Not only are we changing the climate, but human activity is also changing the rate and type of environmental disasters.

Environmental disasters are a 'serious disruption of the functioning of a community or a society at any scale due to hazardous events interacting with conditions of exposure, vulnerability and capacity, leading to one or more of the following: human, material, economic, and environmental losses and impacts' (UNISDR 2017). In April 2021 we reached the tipping point of more than 420 parts per million of carbon dioxide in the atmosphere, which scientists identify as 'half-way to doubling point' of pre-industrial levels (Cappucci and Samnow 2021). Doubling CO2 in the atmosphere is likely to lead to a warming of between 2.3 to 4.5 degrees Celsius. Despite a dramatic decline of CO2 emissions during the global pandemic in 2020, a rebounding of economic activity and a failure of clean energy policies has led to resurgent CO2 emissions (International Energy Agency (IEA) 2021). The current trajectory demonstrates that action to limit greenhouse gas emissions, primarily CO2, but also methane and nitrous oxide, is not happening fast enough to limit catastrophic climate change. This is contributing to feedback effects on other earth systems including two extensive ice melts in late July 2021 of the Greenland Ice Sheet, with more melting expected (Scambos *et al.* 2021). It is also having a multiplier effect on environmental disasters (UNDRR 2019, p. 134). Environmental disasters are recognised as affecting those least likely to be able to withstand them: the vulnerable, and lowest income communities and households in middle and low-income countries. (Park, this issue).

To date, the focus on environmental disasters has been on isolated events. In the northern hemisphere summer in 2021, sudden and unexpected floods in Northern Germany resulted from extreme precipitation; a heat dome over North America caused record-breaking temperatures; and a 7.2 magnitude earthquake hit Haiti in August. Not all environmental disasters are human induced. But they are often exacerbated by the complex inter-relationship between hazards and vulnerabilities, with changing earth systems likely to alter our ability to predict their likelihood (Park, this issue). In response to specific environmental disasters like the global COVID19 pandemic, political leaders scramble to respond immediately, often as though these are exogenous events that interrupt normal activities. Yet, as the IPCC report demonstrates, extreme precipitation and heatwaves are not only on the increase, but they are also becoming more intense, more widespread, and are part of ongoing alterations in Earth systems the effects of which we are unsure (Lenton *et al.* 2019, pp. 592–3). Climate change interacts with the other earth systems: it is one of many drivers that exacerbates habitat loss and species extinction (IPBES 2019, p. 11), which in turn increases the likelihood of the spread of zoonotic diseases and human pandemics, while also increasingly impacting on long term human development outcomes.

This Symposium is not primarily focused on climate change although its hand in environmental disasters is increasingly visible. It seeks to shine the light on political actors and institutions in terms of how they respond to environmental disasters at different scales. We can analyse how political actors frame the accumulation of knowledge and responses to these changes in the earth's systems at the international, national, and local levels. While states try to lock in international agreements to curb emissions through the UNFCCC, various political actors are responding to environmental disasters at the national level through regulations, policies, and action plans, and at the local level where communities must mediate with government and non-government actors such as (multinational) corporations to prevent or mitigate disaster. The symposium seeks to critically examine how actors engage politically with each other and with(in) institutions to address disasters.

The articles in this Symposium examine political responses to environmental disasters according to different scales, timeframes, probabilities, and impacts. Political actors help frame what constitutes the disaster and how it ought to be addressed. This is important because many environmental problems are not decomposable, yet they are often treated as such. Evaluating how actors identify the problem shapes what is and is not addressed. This takes place at all scales. Both the global and the local are important for three reasons: first, because earth systems trends are no longer the background for environmental disasters but actively contribute to them. Clearly some environmental disasters, like floods, heatwaves, and fires must be addressed by the communities affected. But they are increasingly occurring everywhere. We cannot be blind to these trends. Second, because political communities are arranged into units such as sovereign states that delimit what responses to environmental disasters are possible. At the local level this may shape how the problem is mediated through national institutions such as laws, governing bodies, and culture. Finally, because the impact of environmental disasters is felt locally: Communities must respond to wildfire, plagues, and drought. They must also engage with institutions and actors to prevent environmental disasters and manage their ongoing harms, from preventing drilling for fossil fuels, to industrial leaks, oil spills, and mining waste.

The Symposium therefore has three key issues that it seeks to address:

(1) to theoretically interrogate our understanding of what constitutes an environmental disaster and identify the parameters for political responses.
(2) to identify contemporary practices of the state that exacerbate the impact of, and responses to, environmental disasters.
(3) to highlight the strengths and weaknesses in political and institutional responses at the local level to such disasters by state and non-state actors.

In addressing the first issue, the collection provides insight into how different theoretical approaches and concepts can be used to reveal how environmental disasters are understood and what alternative ways of thinking have been overlooked. This means that we can identify how the international framing of responses to environmental disasters contributes to how it has thus far been sidelined in terms of the action required to address environmental disasters. In the first article, I use a constructivist lens to examine how the structure of the sovereign state system limits its ability to address fast moving, systemic, and potentially irreversible change that is increasingly occurring in unanticipated ways. This is because sovereign states have advanced environmental disasters as an external risk to be reduced, outside their collective contribution to both hazards and vulnerabilities. States are autonomous actors with political authority and territorial autonomy, which enables them to extract natural resources for production and destruction. Their behaviour collectively constitutes the international system, and in turn are constituted by it. By ignoring the role of states in contributing to the conditions that enable environmental disasters, they limit the ability to effectively govern increasingly unknowable and cascading environmental disasters.

Contributors to this Symposium move beyond this to identify ways of acting and thinking. Sabine Selchow aims to align the governance of disasters to correspond to a cosmopolitised reality, which is distinctly integrated but also distinctly uncertain. Through a discourse analysis of the United Nations Disaster Risk Reduction (UNDRR) 2019 Global Assessment Report (GAR), she shows openings within the web of meaning of how disasters are governed that recognise how we are all linked within the global political economy, despite our separate political communities. Specifically, she examines how the 2019 GAR invokes the concept of systemic risk, which provides a vehicle for incorporating a broader understanding of 'planetary disaster' for how to think about and therefore govern disasters. Using Ulrich Beck's concept of reflexive modernisation, Selchow demonstrates how systemic risk as a concept highlights how containerised national level decision making around techno-industrial political and economic decisions are reflexively undermined when one brings in concerns for those affected beyond the state, and far into the future, such as those not yet born. The UNDRR as an institution is open and flexible such that an opening for the concept of planetary disasters is possible, which would better reflect a comopolitised reality.

Of course, how disasters are constituted through discourse and framing by states is not the only way to understand our impact on the environment and therefore what ought to be done to mitigate it. Rebecca Lawrence examines how ignorance can be actively produced when assessing the impact of uranium mining. In investigating the prospect of rehabilitating the just closed Ranger Mine in the Northern Territory of Australia in 2021, Lawrence shows how particular scientific knowledge is prioritised to make the process manageable, despite over 1000 leaks from the mine into the waterways and groundwater of a tropical monsoonal environment. This creates uncomfortable knowledge about the toxicity of the mine and the prospect of containing contamination. The ignorance of the entirety of the environmental disaster is actively produced both in terms of acknowledging the immediate impacts of the mine but also its effects deep into the future by the mining corporation, the government regulators, and even the environmental NGOs pushing for rehabilitation in the knowledge that this may not be possible.

The second aim of the Symposium is to investigate contemporary practices of the state that exacerbate the impact of, and responses to, environmental disasters. In other words, it examines different blinkered lenses and limitations in the organisation of the state. The state's complicity in human induced environmental disasters highlights the ways in which energy systems and use are geared towards understanding nature as a resource to be exploited. Christine Winter argues that environmental disasters are unsurprising because they result from an underdeveloped Western liberal philosophy. Drawing on John Locke's understanding of persons with private property and possessions being united with government to create a political community, nature has been understood by the West as something to exploit and improve. Governments extended this principle throughout the world through colonisation. Analysing the High Court challenge to deep sea oil drilling off the coast of New Zealand, Winter uncovers how the government, through carving up separate Departments of Energy from Environment, effectively enable the promotion of environmental disasters while refusing to acknowledge the Maori Te Whānau-ā-Apanui's (Apanui) people's spiritual values in protecting the water, vegetation, and animals of the ocean and sea floor. While the project never came to fruition, it demonstrated the difficulty of the Apanui to challenge the state's drive for exploitation, making potential disasters near a volcano unsurprising.

Third, the Symposium analyses community responses to environmental disasters and how this challenges states' political institutions. As noted, Christine Winter exposes the weaknesses of the New Zealand legal apparatus in ignoring the Maori cosmology connecting people to land and place, which facilitates the potential for environmental disasters in seismic areas. Teresa Kramarz looks across and within states to argue that extractivist disasters, such as those already mentioned, affect the ability of communities in project areas and disaster zones to resist. She argues that we should consider how the temporal dimension of a disaster affects groups, but also to consider the agency of a community in terms of its ability to respond. For example, research into slow violence highlights how community capacity is significantly affected by ongoing exposure to hazards, making groups more vulnerable. Kramarz compares this to fast violence such as the BP oil spill off the coast of Florida, where the immediate impact led to swift responses. Depending on the extent of the disaster and the response, fast occurring disasters can then shift to become slow violence. But we also need to examine communities' capacity to respond to such events. Looking at the vulnerability of the community before a disaster also provides insight into its capacity to react. Here vulnerable subjects may be the very ones we expect: poor, racialized communities in high-risk areas in developing states. But they may also be middle-class communities in industrialized states without the capacity to challenge the perpetrators of slow violence. Kramarz provides a typology of social vulnerability through which we can analyse the capacity of vulnerable subjects to react to environmental disasters.

The 21st century will be more disaster prone. More extreme weather events are likely. Our consumption and production patterns are changing the earth systems (Wiedmann *et al.* 2020, p. 4). Internationally, the United Nations Sustainable Development Goals are a call for action for states to grapple with these environment and development challenges to enable us to live within the Earth's carrying capacity. Multilateral environmental agreements, like the UN Framework Convention on Climate Change and the Sendai framework for Disaster Risk Reduction, require real commitments from states. Meanwhile, an increasing number of environmental defenders are being killed every week for protecting their land and livelihoods against the very extractivism that contributes to environmental disasters (Butt *et al.* 2019, p. 744). Environmental disasters are an environmental issue, a political economy issue, a human rights issue, and a security issue (Scheidel *et al.* 2020, Johnson *et al.* 2021). The politics of 21st century environmental disasters must incorporate the earth's feedback effects into any and all future decision-making.

Disclosure statement

No potential conflict of interest was reported by the author(s).

Funding

This work was supported by the University of Sydney.

ORCID

Susan Park http://orcid.org/0000-0001-7932-1525

References

Butt, N., *et al.*, 2019. The supply chain of violence. *Nature Sustainability*, 2, 742–747. doi:10.1038/s41893-019-0349-4

Cappucci, M. and Samnow, J., 2021. Carbon dioxide spikes to critical record, halfway to doubling preindustrial levels. *The Washington Post* 5 Apr. Available from: https://www.washingtonpost.com/weather/2021/04/05/atmospheric-co2-concentration-record/ [Accessed 17 Aug 2021].

International Energy Agency (IEA), 2021. After steep drop in early 2020, global carbon dioxide emissions have rebounded strongly. Press Release 2, March. Cited: After steep drop in early 2020, global carbon dioxide emissions have rebounded strongly - News - IEA [Accessed 15 July 2021].

IPBES, 2019. Summary for policymakers of the global assessment report on biodiversity and ecosystem services of the intergovernmental science-policy platform on biodiversity and ecosystem services. *In*: S. Díaz, *et al.*, eds. *Global Assessment Report on Biodiversity and Ecosystem Services*. Bonn, Germany: IPBES secretariat, 56. doi:10.5281/zenodo.3553579

IPCC, 2021. Summary for Policymakers. *Climate change 2021: The physical science basis. Contribution of working Group I to the sixth assessment report of the intergovernmental panel on climate change*. Cambridge University Press. In Press.

Johnson, M.F., Rodríguez, L.A., and Quijano Hoyos, M., 2021. Intrastate environmental peacebuilding: a review of the literature. *World Development*, 137, 105150. doi:10.1016/j.worlddev.2020.105150

Lenton, T., *et al.*, 2019. Climate tipping points – too risky to bet against. *Comment, Nature*, 575 (7784), 592–595. doi:10.1038/d41586-019-03595-0

Scambos, T., Stroeve, J., and Koenig, L., 2021. Greenland ice sheet today. National Snow and Ice Data Center, University of Colorado. Available from: http://nsidc.org/greenland-today/ [Accessed 16 Aug 2021].

Scheidel, A., *et al.*, 2020. Environmental conflicts and defenders: a global overview. *Global Environmental Change*, 63, 102104. doi:10.1016/j.gloenvcha.2020.102104

UNDRR, 2019. *Global assessment report on disaster risk reduction*. Geneva, Switzerland: United Nations Office for Disaster Risk Reduction (UNDRR).

UNISDR, 2017. *Terminology*. Geneva, Switzerland: United Nations International Strategy for Disaster Reduction. Online glossary. Available from: https://www.undrr.org/terminology/disaster.

Wiedmann, T., *et al.*, 2020. Scientists warning on affluence. *Nature Communications*, 11 (1), 3107. doi:10.1038/s41467-020-16941-y

The role of the Sovereign state in 21st century environmental disasters

Susan Park

ABSTRACT
2019 was the second hottest year on record, with enhanced ice melts, sea level rises, heatwaves, droughts, and unprecedented large-scale wildfires of record intensity. Here I chart a new research agenda for understanding the interconnections between human behaviour, environmental disasters, and governing a more disaster-prone world. I argue that sovereign states' technological responses to environmental disasters as risks to be reduced systemically ignores human contributions to environmental hazards and vulnerabilities. Using a constructivist analysis, I analyse states' understandings of, and capacity to respond to, cascading environmental disasters at multiple scales according to what they collectively see, what they know, and how they act in terms of emergency preparedness and planning. Revealing the inadequacies of sovereign states' understanding of what constitutes environmental disasters absent their role in them, brings us closer to better governing cascading environmental disasters in the 21st Century.

Introduction

We are in a world of accelerating environmental change that is occurring at a global scale (ESG 2018), with precipitous alterations in natural systems including climate, biosphere integrity, land-system change, and biogeochemical flows (Steffen *et al.* 2015, Lenton *et al.* 2019). This is evidenced by our changing climate affecting enhanced ice melts, sea level rises, heatwaves, droughts, and the unprecedented large-scale wildfires of record intensity that occurred in 2019 (World Meteorological Organisation 2020), with 2019 and 2020 now two of the three warmest years on record (WMO 2020b). The UN Special Representative on Disaster Risk Reduction, Mami Mitzutori, has stated that climate related disasters are occurring at the rate of one per week (Harvey 2019). Human actions are affecting ecological systems, contributing to more environmental disasters (Guha-Sapir *et al.* 2004, IPCC 2018). Environmental disasters are not only more frequent they are occurring with greater severity, which is triggering secondary disasters (AghaKouchak *et al.* 2018). Here I chart a new research agenda for governing a more disaster-prone world using a constructivist lens to understand the interconnections between environmental disasters and sovereign states, as autonomous and purposive actors who primarily bear legitimate authority in the international system (Wendt 1992, Litfin 1998, Pauly and Grande 2007).[1] Specifically, sovereign states have advanced environmental disasters as an external risk to be reduced, outside their collective contribution to both hazards and vulnerabilities, which limit their ability to address fast moving, systemic, and potentially irreversible change that is increasingly occurring in unanticipated ways (UNDRR 2019, p. iv).

My argument proceeds in four steps. First, the increasing likelihood of environmental disasters is outlined, including the types of disasters and how states manage them through preparatory plans, emergency responses, and revising policies to prevent future occurrences. Second, while the interdisciplinary scholarship on disasters includes sociological constructivist approaches (Pantuliano 2020), sovereign states continue to ignore their collective role in initiating disasters and exacerbating hazards and their impacts. To address this, section three argues that we need to go beyond the disaster risk reduction paradigm to examine sovereign states' understandings of and capacity to respond to disasters. Engaging critically with how sovereign states have socially constructed disasters as exogenous risks requires an analysis not only of the nature of cascading disasters at multiple scales but of sovereign states' role in contributing to hazards and vulnerabilities, which result from how disasters are viewed, understood, and acted upon.

A social constructivist approach recognises that 'social agents produce, reproduce and refine the constitutive principles and structures by which and in which they operate' (Litfin 1998, p. 4). While territorially bound, states continue to hold onto the 'organized hypocrisy' of sovereignty that they exist outside the Earth's ecological systems (Krasner 1999, Pauly and Grande 2007). States are autonomous actors with political authority and territorial autonomy, which enables them to extract natural resources for production and destruction (Wendt 1999, p. 53). Their behaviour collectively constitutes the international system, and in turn are constituted by it (Wendt 1992).[2] Although states have ceded some sovereignty to global institutions to address systemic patterns of environmental degradation (Biermann and Dingwerth 2004), efforts to address potentially catastrophic changes in Earth systems through global governance remains fragmented (Biermann *et al.* 2009, Zelli and Van Asselt 2013), mired in accountability traps (Park and Kramarz 2019), and thus far, largely inadequate to address the challenges we face (IPCC 2018, IPBES 2019; WMO 2020). By highlighting how sovereign states externalise disasters as risks to be mitigated rather than acknowledge their contribution to environmental disasters brings us closer to better governing environmental disasters in the 21st Century.

Grappling with more intense, frequent, and cascading environmental disasters

According to the UN Office of Disaster Risk Reduction, 'surprise is the new normal' as global environmental change accelerates such that 'new risks and correlations are emerging in ways that had not been anticipated' (UNDRR 2019, p. iv). This section outlines what constitutes environmental disasters and how they are changing. Environmental hazards are not new. Among them are: earthquakes, volcanic eruption, landslides, floods, tsunamis, extreme precipitation and temperatures, drought and famine, fires, lightening, cyclones, plagues, and epidemics.[3] Some of these hazards occur relatively frequently, like cyclones; others are low probability high consequence events, like volcanic eruptions. Environmental hazards become disasters when they impact on human activity. Disasters are therefore defined as a 'serious disruption of the functioning of a community or a society at any scale due to hazardous events interacting with conditions of exposure, vulnerability and capacity, leading to one or more of the following: human, material, economic, and environmental losses and impacts' (UNISDR 2017).[4] Precisely because of the complex inter-relationship between humans and nature, disasters are no longer understood as purely natural phenomena, occurring absent human intervention (Gaillard 2019).

Previously, disasters were categorised as being caused by an 'act of God' (nature), by humans, or a combination of both. Many hazards are natural but have greater impact because of decisions to build or reside in ecologically fragile or environmentally hazardous areas, such as living in flood plains that experience infrequent but high-volume flooding. This increases the likelihood of damage, harm, and loss (Kelman *et al.* 2015, p. 24). Environmental disasters may also be human induced (often classified as technological). Examples include the largest marine oil spill in history, from the Deepwater Horizon oil rig in the Gulf of Mexico in 2010 (Lawrence 2018), and the world's largest industrial accident in 1984 in Bhopal, India when gas leaked from the Union Carbide pesticide plant killing over 16,000 people. A combination of natural and human factors therefore constitutes an environmental disaster. The impacts of environmental hazards can also be magnified by human activity. An example of a human exacerbated environmental disaster is the 2011 9.0 magnitude earthquake off the coast of Japan that led to the Tohoku Tsunami hitting the Tokyo Electric Power Company's Fukushima Daiichi nuclear reactors. The earthquake and tsunami caused 19,000 casualties, 100,000 people were evacuated from the area when three nuclear reactors went into meltdown, spreading radioactive material.

Classified as a level 7 disaster on the International Nuclear Event Scale Chart (of 1 to 7), this was the worst international commercial nuclear disaster since the 1986 Chernobyl incident in Ukraine (Ismail-Zadeh *et al.* 2014).

Human activity can therefore initiate or exacerbate rapid disasters but can also contribute to slow onset disasters. Humans are mismanaging natural systems including land, biodiversity, and climate contributing to mass species extinction, and soil degradation and salinization, among other impacts. Indeed, a 'major driver of disaster losses and failed development' is the changing climate (UNDRR 2019, pp. iv, x). This means that the changing climate 'drives both hazards and vulnerabilities' (Kelman *et al.* 2015, p. 25). Since 1980, 'each decade has been warmer than any preceding decade since 1850' (WMO 2020). The IPCC (2018) reports that '[H]uman activities are estimated to have caused approximately 1.0°C of global warming above pre-industrial levels' and that trends in intensity and frequency of some climate and weather extremes have been detected over time spans during which about 0.5°C of global warming has occurred. 2019 had a 'global average temperature of 1.1°C above estimated pre-industrial levels' along with enhanced ice melts, sea level rises, and heatwaves, droughts, and unprecedented largescale wildfires of record intensity. Human induced global warming is facilitating slow onset disasters including rising sea levels, ocean warming, and acidification. More worryingly, the IPCC highlights that member state's Nationally Determined Commitments are on track to realise a warming of 2.9 to 3.4 degree warming world (IPCC 2018), resulting in 'future hydrometeorological hazard intensities that surpass known experience and which alter loss and damage equations and fragility curves of almost all human and natural systems at risk' (UNDRR 2019, p. 32).

Since 1960 there has been a significant increase in the frequency of environmental disasters like extreme hydrological and meteorological events, even with variation in data collection over time (Cutter *et al.* 2015, EMDAT 2020). Moreover, environmental disasters are increasingly not isolated events that enable discrete responses. Coupled events, like heatwaves and droughts, are on the increase. Cascading disasters are where environmental disasters occur more frequently and with greater severity triggering secondary disasters. Multiple hazardous events can also cascade to become large-scale disasters. For example, record temperatures and prolonged drought contribute to dry soils and vegetation that increase the likelihood of fires. The barren landscape in the aftermath of fires is then more vulnerable to flooding and mudslides (AghaKouchak *et al.* 2018).

In 2019, wildfires burned in the Arctic, the Amazon, Australia, Indonesia, Siberia, and the US. More than 100 fires raged in the Artic, releasing 50 mega-tonnes of CO2 in June 2019, the equivalent of the previous eight Junes combined. Such sustained burning impacts the weather and climate (NASA 2019). Australia experienced over 200 catastrophic fires over the 2019–2020

'Black Summer,' and fire chiefs stated that the length, extent, and intensity of the bush fires was unprecedented (Readfearn 2019). The ferocity of the bushfires included unmanageable firestorms in parts of Australia that had never previously experienced them (Di Virgilio *et al.* 2019). We know that the natural cycle for bush replenishment will be impeded by the increased likelihood of another earlier and longer fire season. Such events create feedback loops, including further adding CO2 to the atmosphere, for further cascading disasters. Scientists have pointed out that while we tend to study the likelihood of separate disasters such as droughts or floods, we have not yet begun to examine them in concert, or how one triggers the other and compounds their impact (Ismail-Zadeh *et al.* 2014, AghaKouchak *et al.* 2018).

Economic losses from extreme weather events have increased by 251% since the late 1970s (Guha-Sapir *et al.* 2004, UNDRR 2018, p. 5).[5] Most environmental disaster impacts are highly unequal, with those most affected – primarily by high frequency, low consequence events – in middle and low-income states who bear the greatest relative costs and have the least capacity to respond (UNDRR 2019, pp. vi, 29, 278). This is because of the intersection between hazard, exposure, and vulnerability. Across all states, disasters tend affect the most vulnerable; the lowest income communities and households (UNDRR 2019). The good news is that there has also been a significant decline in the number of deaths from environmental disasters globally, from over one million per annum in the mid-twentieth century, to less than 20,000 per annum in the last few decades. This is because of improvements in infrastructure and disaster prediction and response systems (Ritchie 2014, EM-DAT 2020).

Such improvements may be offset by the increasing likelihood of more frequent, intense, and cascading disasters resulting from human activity. Indeed, 'disasters will continue to be major impediments to sustainable development so long as the economic incentives to build and develop hazard-prone locations outweigh the perceived disaster risks' (UNDRR 2018, p. 5). Although previous understandings of sovereignty held that political autonomy of states was absolute, states have agreed to make sovereignty contingent on the protection of people through the Responsibility to Protect norm (Biersteker 2013). Increasingly, people are demanding that states be held responsible for meeting their international climate commitments to reduce the catastrophic effects of climate change including environmental disasters (Kaminski 2019).

However, the current approach to engaging with environmental disasters remains essentially reactive, as sovereign states continue to provide more resources for international disaster relief than for prevention (IFRC and UNDP 2014, UNDRR 2019). Nevertheless, there has emerged a shared global framework for managing disasters and reducing risk with the UN Office of

Disaster Risk Reduction (UNDRR) at the centre. The office was created at the culmination of the UN's International Decade for Natural Disaster Reduction (1990–1999). The UNDRR focus is to calculate and mitigate risk from disasters, build resilience, and achieve the Sustainable Development Goals through the Sendai Framework for Disaster Risk Reduction, 2015–2013 (UNISDR 2015). This integrative framework moves away from hazard-based disaster management to adopt risk as a technology to grapple with human vulnerability and the uncertainty of the increasing frequency, intensity, and systemic nature of disasters (see Selchow, this issue). Both the Sendai Framework and the scholarly literature view disasters as risks to be managed, which I discuss next.

Understanding environmental disasters as reducible risks

According to the UN Office for Disaster Risk Reduction, the 'era of hazard-by-hazard risk reduction is over.' Given the super-charged nature of environmental disasters that are no longer linear and isolated, we need to identify how sovereign states can address unanticipated, fast moving, systemic, and potentially irreversible environmental change (UNDRR 2019, p. iv). This section looks at how the interdisciplinary disaster literature emerged and how it understands disasters as risks to be reduced, which informs states' efforts to govern disasters globally.[6] As I argue below, this delimits our understanding of disasters to probabilistic calculation of a likely event, which may not be possible given that we are now experiencing accelerating change in Earth systems at multiple scales.

The scholarly literature has been wrestling with how to best address environmental disasters for decades. In the 1970s, scholars challenged the view that disasters were natural (the hazard paradigm), to argue that environmental hazards impacted people that were exposed and vulnerable because of cultural, economic, political, and social factors (Ball 1975, O'Keefe *et al.* 1976). The critique of the hazard paradigm was that it focused on an external agent (nature) acting on impassive victims, through which a scientific and technical response was necessary. The vulnerability turn in disaster scholarship in the 1980s sought instead to focus on the social dimension of disasters, highlighting the inter-play between forces that lock the vulnerable into disaster prone areas. The literature highlighted how natural conditions were being exacerbated by exploitative patterns of international development and ecological mismanagement (Gaillard 2019, p. S10).

Over time the literature on disasters would examine the concept of vulnerability, which in turn emphasised human resilience, adaptability, and risk aversion.[7] By the 1990s the concept of vulnerability was being incorporated into the emerging international disaster reduction framework. Critics argued that vulnerability was being 'recast as yet another of the "social pathologies" like, or derived from poverty, underdevelopment, overpopulation. Any problem or other setting that lacks "advanced" technical and governmental organisation, and a strong consumer economy is seem to be prone to these pathologies' (Hewitt 1995, p. 325). Since then, the literature has better integrated how vulnerability stems from both natural hazards and socio-economic causes in its efforts to prevent, mitigate, adapt, and reduce the risk of disasters (Adger 2006, p. 271, Young *et al.* 2006). Recognising the complex inter-play of natural, political, and structural causes of disasters, the interdisciplinary scholarship has framed disasters as risks that can be reduced (Eiser *et al.* 2012), which in turn has become central to states collective efforts to address disasters, as evidenced by the UN Sendai Framework (2015).

In promoting the Sendai Framework, the UNDRR and other international organisations like the Association of South East Asian Nations (ASEAN) have taken up the idea of minimising the 'risk' of disasters (Selchow, this issue; Di Mauro 2014, Pennisi Di Floristella 2016, UNDRR 2019, p. 32). Risk is a technological device to manage disaster uncertainty, and of modernity more broadly, 'which has left the world more exposed to a range of unforeseen, unintended and unknowable outcomes' (Beck 1992, Pennisi Di Floristella 2016, p. 286). The UNDRR supports states through applying the risk reduction framework using probabilistic techniques to assess and forecast the likelihood of events and associated losses (Di Mauro 2014). This can work for predicting and forecasting specific hazards, particularly those that are of high frequency like typhoons. It is much harder for low probability events based on infrequent occurrences, such as big magnitude earthquakes, as it is more difficult for scientists to identify the distribution points for analysis (Eiser *et al.* 2012, p. 7).

However, in understanding environmental disasters as risk, states 'discount the importance of uncertainty' (Blyth 2002, p. 8). Uncertainty is broader than risk, because it is 'not limited to the ability to make a probabilistic calculation regarding the likelihood of some event ... [R] ather, it is a more pervasive constraint, one which limits the ability of agents to form any meaningful estimate of future trends' (Widmaier *et al.* 2007, p. 750). I argue that using risk as a technological device to manage uncertainty delimits states' ability to respond to the changing nature of environmental disasters in the 21st century. Foremost, this is because states have enabled the altering of the Earth's systems in unforeseen ways. This alters the background conditions for baseline studies (like the climate), which changes the ability to probabilistically calculate the likelihood of an event (deductively), shifting forecasting towards estimations of likelihood (inductively). Estimating the likelihood of an event, in turn, relies on a host of assumptions in interpreting the information at hand. This becomes more complicated because rapidly changing Earth systems are now producing aleatory uncertainty (or, uncertainty that is intrinsic to the thing itself, Eiser *et al.* 2012, p. 7), through creating new forms of unanticipated, non-linear, multi-hazard disasters. The use of risk as a heuristic device helps states and international organisations manage environmental disasters. Yet I argue below that this does not go far enough because it ultimately does not recognise that disasters are not exogenous and isolated events beyond the states' responsibility. Rather, environmental disasters are endogenous to the social foundations of the sovereign state system, with implications for the insecurity of states in a more disaster-prone world.

Sovereign states' collective limitations in seeing, knowing, and acting on environmental disasters

How can we govern more frequent and more severe cascading environmental disasters? I argue that a major stumbling block to addressing the increasing frequency, intensity, and systemic nature of disasters is the social construction of the sovereign states as separate from the Earth's planetary boundaries. This delimits system-level responses to environmental disasters to collective action games according to who benefits the most from international cooperation. These take place either through negotiations in multilateral environmental agreements on specific drivers of environmental disasters, like climate change through the UNFCCC, or voluntarily adhering to the UNDRR guidelines. This precludes viewing states' activities as systemically contributing to both hazards and vulnerabilities in a more disaster-prone world.

Wisner highlights that the United Nations has helped galvanise technical knowledge, support for building institutions, and for the provision of grants and loans to assist states in disaster risk reduction, but it has not been able to build a 'moral imperative that can mobilise local political will' (2001, p. 126). Given the fluidity between the changing nature of Earth systems and the multiple scales at which environmental disasters are occurring, states may need to go beyond the methodological nationalism they collectively endorsed in the UNDRR (see Selchow this issue). In contrast, a methodological holist approach (Finnemore and Sikkink 2001, p. 393), which analyses how states constitute environmental disasters as exogenous risks to be managed, reveals three limitations: First, the social construction of sovereign states as politically autonomous and independent units with control over a containerised territory sealed off from ecological systems means that states do not *see* disasters as normal and reoccurring, enabling them to constantly discount their likelihood and impact. Second, states remain vulnerable because they *understand* and therefore prioritise governing economic and political activities over knowledge of the relationship between disasters and (failures of) sustainability; and third, as a result of not seeing and understanding environmental disasters, states overestimate their institutional capacity to *act* in response to perceived disasters and underestimate societies' reactions to early warning systems. These limitations contribute to the likelihood of greater hazards and vulnerabilities for the sovereign state system as a whole, with implications for governing disasters in the 21st century. Each is addressed in turn.

Seeing environmental disasters

Although non-state actors including Multinational Corporations and transnational activists play important roles in contributing to hazards and addressing vulnerabilities, states are the only actor with the authority and capacity to reshape global social-ecological relations in ways that normalise rapid and slow onset disasters within state functions (Eckersley 2004, Young *et al.* 2006). Yet states do not see disasters as endogenous because they are not considered normal. Disasters are 'exogenous, unexpected, extreme events that randomly affect otherwise "normally" developing societies' (Chipangura *et al.* 2019, p. 5). States' regulatory responses may vary according to 'myths of nature and normative views on how to react to exogenous shocks' (Parrado 2018, p. 3). Precisely because disasters are exogenous, states underestimate the likelihood of disasters and discount their impact (UNDRR 2019, p. 36). This in part rests on access to and use of hazards science: for example, seismic research is based on the probability of an earthquake occurring, but extreme seismic events that took place throughout the 2000s including the 2004 Sumatra-Andaman, 2005 Kashmir, 2008 Wenchuan, 2010 Chile, and 2011 Tohoku earthquakes were 'big magnitude events [that] c[a]me as a surprise, even though scientific knowledge of historical large earthquakes was available in the most of these areas. Moreover, existing seismic hazard maps did not mark these regions as those of significant hazards.' Fatalities from these events were underestimated by a magnitude of two-to-three times (Ismail-Zadeh 2014, p. 49). Although there remains imprecision in hazards science, disasters can be reduced with better infrastructure and greater disaster prediction and response systems (Ritchie 2014). States therefore rely on improving scientific knowledge to reduce the economic, human, and environmental risks stemming from environmental disasters.

However, it is not just the uncertainty of the likelihood of environmental hazards and therefore estimations of their occurrence and impacts. It also speaks to state's perpetuation of economic growth and development patterns that exclude environmental factors, in ways that directly contribute to environmental disasters (Kelman *et al.* 2015, p. 25). For example, unsustainable development practices contributed to large-scale deforestation in China increasing the scale of floods in 1998, and the removal of wetlands in the US contributing to more severe tropical storms and hurricanes. By not seeing disasters and discounting their impacts, states do not need to reconcile their role in contributing to environmental disasters, even when prevention is economically cheaper than reconstruction (Cutter *et al.* 2008, p. 601, 2015). Moreover, the further into the past a disaster becomes, the more likely states will discount its recurrence (see for example, US efforts to rollback regulating oil spills after Deepwater Horizon, Boesch 2020). This highlights an underestimation of the risks entailed and an overestimation of being able to manage environmental disasters.

For some, the state will not be able to obtain a socio-ecological transformation because there is an inherent glass ceiling within the structure of the state itself, which is based on a global unsustainable economic growth model of accumulation and consumption (Hausknost 2020, p. 19). This prevents advanced industrialised states shifting from their current position as 'environmental states,' with institutions and practices to manage societal-environment interactions (Duit *et al.* 2016, pp. 5–6), like disaster risk reduction, to transforming into 'green states' that prefer sustainability over economic growth (Eckersley 2004).[8] This is especially so for neoliberal environmental states like the United Kingdom, that prefer the financial system over other public goods and where environmental institutions have to conform to the economic rationality of the free market (Hatzisavvidou 2020).[9] Being plugged into the global economy means that developing states have been encouraged to emulate the 'environmental state' path, reinforcing the myopia that disasters are exogenous and unexpected for normally developing societies and that their risks can be managed. Others point to the developmental state model as a green pathway (Kim and Thurbon 2015), although this is still predicated on export-oriented economic growth model of accumulation and consumption. States therefore continue to promote economic growth, patterns of development, and globalisation that contribute to environmental degradation (Smith 2009, pp. 99–100), leading to ecological feedback loops that reinforce the likelihood of more intense, frequent, and coupled hazards leading to more cascading disasters.

However, disasters are now being connected to larger environmental and development aims including the Sustainable Development Goals, climate change adaptation, and provision of international development aid (Shipper and Pelling 2006, Islam *et al.* 2020). States have constructed global environmental norms to address environmental challenges to moderate economic growth (Warner 2005, Okereke 2008). States uptake of the concept of sustainable development in the 1990s aimed to reconcile economic growth and environmental degradation (WCED 1987), through 'liberal environmentalism' (Bernstein 2001). Scholars therefore point to the potential for a reckoning, with slow onset environmental disasters like climate change challenging the 'larger cultural complex of modernity' in which the sovereign state is bound, because the environment has grassroots constituents, is underpinned by science, may cross borders, and has an intergenerational temporal component (Litfin 1998, p. 16, Eckersley 2004). What this highlights is the potential for states to reframe their understanding of environmental disasters as normal socio-ecological events that they are responsible for, a point further developed below.

Knowing environmental disasters

States' reactive response to environmental disasters (IFRC and UNDP 2014, UNDRR 2019, p. 319) has led to a 'call for a shift from ad hoc, disaster-driven, and reactive systems and policies to a proactive, threat-driven, and mitigative focus' (Cutter *et al.* 2008, p. 601). As mentioned above, states face significant impediments in addressing environmental disasters, including the ability to collect and monitor data and implement disaster prevention and mitigation frameworks nationally and sub-nationally (UNDRR 2019). However, it may not just be states' incapacity that contributes to ongoing failures to address environmental disasters. For example, '[D]espite nearly a half century of concerted research and public policy practice, the US government remains uncommitted to reducing society's vulnerability to natural hazards' (Cutter *et al.* 2008, p. 598). This suggests that states' disaster responses may be political and economic in determining how much risk a state is willing to bear.

To adopt a mitigative focus on environmental disasters, sovereign states would need to normalise hazards and vulnerabilities. This would mean a reorientation of the sovereign state, beyond territory as constituting political autonomy over a given area, non-interference, and the Permanent Right over Natural Resources (Biersteker 2013, pp. 256–66). State sovereignty is not absolute, it can and does change in accordance with prevailing norms over time. Indeed, previously hard borders are now known to be porous as flows of capital, goods, people, and transnational environmental impacts attest (Krasner 1999). The Responsibility to Protect norm also demonstrates that non-interference, which is enshrined in the United Nations Charter, is contingent on states willingness and ability to protect their own people. Likewise, the right to exploit natural resources, articulated by states in both the Stockholm (Principle 21) and Rio Declarations (Principle 2), has increasingly been moderated to 'respect the rights and interests of indigenous peoples, and a duty to use natural wealth and resources in a sustainable way' (Schrijver 2010, pp. 7–8). To address the feedback loop between human activity and systemic environmental disasters, sovereign states would need to further strengthen their collective obligations to the Earth's systems including land use, biodiversity, and climate, among others.

The urgency of the problem means that sovereign states must internalise environmental disasters domestically as well. This challenges states' political structures: political leaders are not often rewarded for events that did not happen (averting disasters). Disaster preparation and planning may not be rewarded within the timeframe of a democratic election cycle (Cutter *et al.* 2015, French *et al.* 2020, p. 11), nor do disasters tend to trigger change in stable authoritarian states (Aldrich *et al.* 2019). State leaders may prioritise short-term political and economic concerns, which may include not enforcing disaster averting regulations, practising emergency preparation plans, using disaster recovery funds for future mitigation efforts (Cutter *et al.* 2015, French *et al.* 2020). Yet there are attempts to overcome political and economic limitations: there are pushes for states to 'build back better' nationally to ensure mitigation is part of any recovery process, and internationally there are attempts to bring together knowledge and practitioners from development, disaster risk reduction, and climate change adaptation (UNISDR 2015). Scholars highlight that ensuring sustainable ideas throughout state bureaucracies remains problematic (Heinrichs and Laws 2014), while environmental disasters tend to be the isolated response of one government department or agency despite multiple hazards and cascading disasters (Kachali *et al.* 2018).[10] Recognisably, there are broader epistemological difficulties in bringing together different disciplinary knowledges to tackle the multifaceted dimensions of normalising environmental disasters (Shipper and Pelling 2006, Ismail-Zadeh *et al.* 2014). This directly informs how states respond to environmental disasters, as discussed next.

Acting on environmental disasters

Sovereign states view environmental disasters as exogenous events that can be managed within a risk framework, rather than viewing hazards and vulnerabilities as endogenous to state sovereignty. This contributes to failures to anticipate environmental disasters as noted above, and to adequately respond to them politically, economically, and as I argue here, socially, and culturally. Without internalising hazards and vulnerabilities, states' emergency preparations and planning may be inadequate for future disasters in two ways: first, states may overestimate their institutional capacity to enact emergency preparedness plans in response to perceived disasters, and second, states may presume that early warning systems will be followed as intended. First, states may overestimate their capacity to respond to perceived disasters, which is itself premised on two components: first, that states tend to prepare for disasters based on previous experience, but where previous disasters may not be replicated. Quite simply, states tend to prepare for past experiences and seek to prevent their reoccurrence (see French *et al.* 2020). As discussed earlier, future environmental disasters are likely to be qualitatively different than previous ones. In this regard, stability in emergency preparations and planning in being able to meet previous disasters can lead to overconfidence that can lead to an inability to address future (unanticipated) disasters (Widmaier 2016).

Second, states may overestimate their ability to enact an emergency preparedness plan and underestimate the impacts of the disaster. States may not have experience in enacting their plans leading to failures to adequately address the environmental disaster (Cutter *et al.* 2015). Without practice, states may not know whether the plans they have prepared work. This is because any planning or 'production process depends on a host of informal practices and improvisations that could never be codified' (Scott 1998, p. 6). The current COVID-19 pandemic is illustrative in revealing three weaknesses in many advanced European states' emergency preparation plans: first, because they failed to enact their plan; second, they failed to ensure sufficient national stocks of emergency equipment (in this case diagnostic testing kits, personal, protective equipment, ventilators, and oxygen concentrators). Both may have results from overconfidence that they were adequately prepared because they had state capacity (Kirkpatrick *et al.* 2020). And third, a failure to recognise changing background conditions that had not been incorporated into the emergency preparedness plan. For example, European states relied on global supply chains of medical equipment from China for a pandemic outbreak, which were also interrupted with factories shutting down, delays in production and transportation, and an increase in panic buying (UN 2020).[11]

Finally, even if states have emergency preparation plans in place, the disaster risk reduction approach is built on a managerial technical assumption that society will respond to Early Warning System announcements as states intend, such as warnings to evacuate before an environmental hazard (Bankoff 2003). However, there are social and cultural reasons for people to discount the risk and ignore state warnings. This may include for example mistrust in government, meaning that people do not believe state announcements. It may include a misunderstanding of the risk being conveyed because of the technical terms being used (Bankoff 2003). Or it may also stem from cultural beliefs such as believing that disasters are an act of god, which make evacuation seem unnecessary (Ayeb-Karlsson *et al.* 2019). Research shows that communities are able to better adapt to consecutive disasters where there is trust in the government and local and indigenous knowledge is used to convey information (Ching *et al.* 2020). In many ways this challenges the risk reduction framework that relies on Western epistemologies that ignore local context and indigenous knowledge (Bankoff 2003, Gaillard 2019; see Winter, this issue).[12] According to Adger 'policy interventions need to recognize the plurality of types of knowledge and of governance systems that are used throughout the world to manage risks and promote resilience'

(2006, p. 276). Thus, even states like Peru that have strong disaster risk management preparations, and are used to frequent disasters, recognise that they may be ineffective in minimising environmental disasters if they continue to apply homogenous international frameworks absent local institutional contexts (French *et al.* 2020, p. 10). Additionally, small scale localised approaches may also provide communities with stronger adaptive capacities to environmental disasters (Freeman and Hancock 2017). This highlights how states can better govern environmental disasters: recognise their own contribution to disasters, incorporate multi-disciplinary knowledge of disasters, and incorporate localised understandings of disasters and adaptive strategies to respond to them (see for example Millar and Pescaroli 2018, Pescaroli 2018). How to govern future environmental disasters therefore relies significantly on both external sovereignty in terms of states collectively recognising their role in contributing to both hazards and vulnerabilities, and internal sovereignty in terms of protecting their people and engaging with localised understandings of disasters (Pauly and Grande 2007).

Conclusion

Although non-state actors such as corporations and environmental non-government organisations can be legitimate bearers of authority, 'collective responsibility for the health of the planet,' including reacting to rapid disasters like pandemics and slow onset disasters like climate change, remain the domain of sovereign states (Litfin 1998, p. 16). Here I have highlighted the weaknesses of sovereign states in responding to a more disaster-prone world. I identified the increasing likelihood of environmental disasters in the twenty-first century that will surpass known experience, irrevocably changing human and natural systems. This fundamentally alters our understanding of the relationship between humans and the biosphere, and what it means to govern environmental disasters. To address more intense, severe, and systemic environmental disasters, sovereign states must go beyond the disaster risk reduction paradigm. States have socially constructed disasters not as normal and endogenous to state practices but as exogenous risks that requires managerial solutions. This leaves states vulnerable to cascading disasters because they do not see them as normal ecological responses to human activity, such that they therefore discount their likelihood and impact. States continue to therefore prioritise political and economic activities over knowledge of the relationship between disasters and (failures of) sustainability. Moreover, states overestimate their capacity to respond to perceived disasters, as evidenced by the global COVID-19 pandemic, and states continue to underestimate peoples' reactions to top-down Early Warning Systems. Revealing the inadequacies of sovereign states' understanding of their role in contributing to hazards and vulnerabilities demonstrates their inability to address them. Highlighting the connections between state sovereignty and environmental disasters brings us closer to better governing disasters in the 21st Century.

Notes

1. Sovereign states of all political and economic regime types and levels of development (democratic, authoritarian, liberal/neoliberal/welfare capitalist, communist, developed and developing) contribute to, and must address, environmental disasters. For this reason, the capacity and policy autonomy of specific types of states to respond to environmental disasters is not detailed here, although various institutionalist approaches in political science have been usefully applied (e.g. Lin 2015, Valdivieso and Andersson 2017).
2. A constructivist approach differs from the main international relations theories, that either argue that states have too little power to effect change in the international system (neorealism and neoliberal institutionalism) or that states can effect change on behalf of domestic social forces (liberal internationalism, Marxist, poststructuralist, and neo-Gramscian approaches). See Hobson (2002) and Carroll *et al.* (2020). Constructivism looks at how states constitute the international system, and through their collective understanding and actions, can change it (Wendt 1992).
3. The International Disaster Database distinguishes natural disasters by class: geophysical, meteorological, hydrological, climatological, biological, and extra-terrestrial, compared with (human made) technological disasters: industrial, transport, and miscellaneous. A disaster must meet the following criteria to be entered into the database: 'Ten or more people reported killed; 100 or more people reported affected; Declaration of a State of Emergency; Call for international assistance' (EM-DAT 2020).
4. Vulnerability is defined as the 'state of susceptibility to harm from exposure to stresses associated with environmental and social change and from the absence of capacity to adapt' (Adger 2006). The literature differs on the relationship between vulnerability and exposure to a hazard (Cutter *et al.* 2008, p. 600).
5. This is a result of better reporting of assets (UNDRR 2019, p. 269), economic and population growth, and building expansion (Ritchie 2014).
6. This is embodied, for example, in the *International Journal of Disaster Risk Reduction*, which launched in 2012.
7. Resilience is defined as the 'ability of a social system to respond and recover from disasters and includes those inherent conditions that allow the system to absorb impacts and cope with an event, as well as post-event, adaptive processes that facilitate the ability of the social system to re-organize, change, and learn in response to a threat' (Folke 2006, Cutter *et al.* 2008, p. 599).
8. See also Meadowcroft (2005) and Heinrichs and Laws (2014).
9. A range of international financial services have been identified as necessary to finance environmental disaster risk reduction (see UNDRR 2019).
10. Scholars also highlight a different response is needed to protect critical infrastructure from environmental disasters (Feng and Xiang-Yang 2018).

11. States' reliance on artificial intelligence and networked information may also provide another vulnerability point in responding to environmental disasters (UNDRR 2019).
12. Although there is recognition of traditional indigenous knowledge in the UNDRR report (2019).

Disclosure statement

No potential conflict of interest was reported by the author(s).

References

Adger, W.N., 2006. Vulnerability. *Global Environmental Change*, 16, 268–281. doi:10.1016/j.gloenvcha.2006.02.006

AghaKouchak, A., *et al.*, 2018. How do natural hazards cascade to cause disasters? Comment. *Nature*, 561, 458–460. doi:10.1038/d41586-018-06783-6

Aldrich, D., Forrester, S., and Horhager, E., 2019. Triggers for policy change: the 3.11 Fukushima meltdowns and nuclear policy continuity. *Environmental Politics*, 28 (7), 1214–1235. doi:10.1080/09644016.2018.1510216

Ayeb-Karlsson, A., *et al.*, 2019. I will not go, I cannot go: cultural and social limitations of disaster preparedness in Asia, Africa, and Oceania. *Disasters*, 43 (4), 752–770. doi:10.1111/disa.12404

Ball, N., 1975. The myth of the natural disaster. *The Ecologist*, 5 (10), 368–371.

Bankoff, G., 2003. *Cultures of disaster: society of natural hazard in the Philippines*. New York and London: Routledge.

Beck, U., 1992. *Risk society. Towards a new modernity*. London: Sage.

Bernstein, S., 2001. *The compromise of liberal environmentalism*. New York: Columbia University Press.

Biermann, F., *et al.*, 2009. The fragmentation of global governance architectures: a framework for analysis. *Global Environmental Politics*, 9 (4), 14–40. doi:10.1162/glep.2009.9.4.14

Biermann, F. and Dingwerth, K., 2004. Global environmental change and the nation state. *Global Environmental Politics*, 4 (1), 1–22. doi:10.1162/152638004773730185

Biersteker, T., 2013. State, Sovereignty and territory. *In*: W. Carlsnaes, T. Risse, and B. Simmons, eds. *Handbook of international relations*. Los Angeles: Sage, 245–272.

Blyth, M., 2002. *Great transformations: economic ideas and institutional change in the twentieth century*. Cambridge: Cambridge University Press.

Boesch, D., 20 April 2020. *A decade after the deepwater horizon explosion, offshore drilling is still unsafe*. The Conversation, Australian Edition. Available from: https://theconversation.com/a-decade-after-the-deepwater-horizon-explosion-offshore-drilling-is-still-unsafe-132846 [Accessed 28 July 2020].

Carroll, T., Hameiri, S., and Jones, L., eds., 2020. *The political economy of South East Asia*. 4th ed. UK: Palgrave Macmillan.

Ching, A., Morrison, L., and Kelley, M., 2020. Living with natural hazards: tropical storms, lava flows and the resilience of island residents. *International Journal of Disaster Risk Reduction*, 47, 101546. doi:10.1016/j.ijdrr.2020.101546

Chipangura, P., Van Niekerk, D., and Van Der Waldt, G., 2019. An exploration of the tractability of the objectivist frame of disaster risk in policy implementation in

Zimbabwe. *Jàmbá: Journal of Disaster Risk Studies*, 11 (1), a604. doi:10.4102/jamba.v11i1.604

Cutter, S., *et al.*, 2008. A place-based model for understanding community resilience to natural disasters. *Global Environmental Change*, 18, 598–606. doi:10.1016/j.gloenvcha.2008.07.013

Cutter, S., *et al.*, 2015. Global risks: pool knowledge to stem losses from disasters Comment. *Nature*, 522, 277–279. doi:10.1038/522277a

Di Mauro, M., 2014. Special issue: probabilistic assessment of risk from natural hazards at global scale. *International Journal of Disaster Risk Reduction*, 10, 403–405. doi:10.1016/j.ijdrr.2014.11.007

Di Virgilio, G., *et al.*, 2019. Climate change increases the potential for extreme wildfires. *Geophysical Research Letters*, 46, 8517–8526. doi:10.1029/2019GL083699

Duit, A., Feindt, P.H., and Meadowcroft, J., 2016. Greening leviathan: the rise of the environmental state? *Environmental Politics*, 25 (1), 1–23. doi:10.1080/09644016.2015.1085218

Earth Systems Governance Project, 2018. *Science and implementation program of the Earth Systems Governance Project*. Utrecht, The Netherlands: Earth Systems Governance Project.

Eckersley, R., 2004. *The green state. rethinking democracy and Sovereignty*. Cambridge, MA: MIT Press.

Eiser, J., *et al.*, 2012. Risk interpretation and action: a conceptual framework for responses to natural hazards. *International Journal of Disaster Risk Reduction*, 1 (1), 5–16. doi:10.1016/j.ijdrr.2012.05.002

EMDAT, 2020. *OFDA/CRED international disaster database*, Université catholique de Louvain – Brussels – Belgium. Available from: https://ourworldindata.org/natural-disasters#wildfires [Accessed 25 June 2020].

Feng, Y. and Xiang-Yang, L., 2018. Improving emergency response to cascading disasters: applying case-based reasoning towards urban critical infrastructure. *International Journal of Disaster Risk Reduction*, 30, 244–256. doi:10.1016/j.ijdrr.2018.04.012

Finnemore, M. and Sikkink, K., 2001. Taking stock: the constructivist research program in international relations and comparative politics. *Annual Review of Political Science*, 4, 391–416. doi:10.1146/annurev.polisci.4.1.391

Folke, C., 2006. Resilience: the emergence of a perspective for social–ecological systems analyses. *Global Environmental Change*, 16, 253–267. doi:10.1016/j.gloenvcha.2006.04.002

Freeman, J. and Hancock, L., 2017. Energy and communication infrastructure for disaster resilience in rural and regional Australia. *Regional Studies*, 51 (6), 933–944. doi:10.1080/00343404.2016.1146403

French, A., *et al.*, 2020. Root causes of recurrent catastrophe: the political ecology of El-Nino-related disasters in Peru. *International Journal of Disaster Risk Reduction* 47(101539):1–14

Gaillard, J.C., 2019. Disaster studies inside out. *Disasters*, 43 (S1), S7–S17. doi:10.1111/disa.12323

Guha-Sapir, D., Hargitt, D., and Hoyois, P., 2004. *Thirty years of natural disasters 1974-2003: the numbers*. Belgium: Centre for Research on the Epidemiology of Disasters, Presses Universitaires de Louvain.

Harvey, L., 2019. One climate crisis disaster happening every week, UN warns. *The Guardian*, Sunday 7 July. Available from: https://www.theguardian.com/environment/2019/jul/07/one-climate-crisis-disaster-happening-every-week-un-warns [Accessed 12 June 2020].

Hatzisavvidou, S., 2020. Inventing the environmental state: neoliberal common sense and the limits to transformation. *Environmental Politics*, 29 (1), 96–114. doi:10.1080/09644016.2019.1684732

Hausknost, D., 2020. The environmental state and the glass ceiling of transformation. *Environmental Politics*, 29 (1), 17–37. doi:10.1080/09644016.2019.1680062

Heinrichs, H. and Laws, N., 2014. "Sustainability state" in the making? Institutionalization of sustainability in German federal policy making. *Sustainability*, 6 (5), 2623–2641. doi:10.3390/su6052623

Hewitt, K., 1995. Excluded perspectives in the social construction of disaster. *International Journal of Mass Emergencies and Disasters*, 13 (3), 317–339.

Hobson, J., 2000. *The state and international relations*. Cambridge: Cambridge University Press.

IFRC and UNDP, 2014. *Effective law and regulation for disaster risk reduction: a multi country report*. New York: International Federation of the Red Cross, and the United Nations Development Programme.

IPBES, 2019. *Summary for policymakers of the global assessment report on biodiversity and ecosystem services of the intergovernmental science-policy platform on biodiversity and ecosystem services*. Bonn: IPBES.

IPCC, 2018. *Global warming of 1.5°C*. Geneva: IPCC, An IPCC Special Report.

Islam, S., Chu, C., and Smart, J., 2020. Challenges in integrating disaster risk reduction and climate change adaptation: exploring the Bangladesh case. *International Journal of Disaster Risk Reduction*, 47, 1010540. doi:10.1016/j.ijdrr.2020.101540

Ismail-Zadeh, A., *et al.*, eds., 2014. *Extreme natural hazards, disaster risks and societal implications*. Cambridge: Cambridge University Press.

Ismail-Zadeh, A., 2014. Extreme seismic events: from basic science to disaster risk mitigation. *In*: A. Ismail-Zadeh, *et al.*, eds. *Extreme natural hazards, disaster risks and societal implications*. Cambridge: Cambridge University Press, 47–60.

Kachali, H., *et al.*, 2018. Inter-sectoral preparedness and mitigation for networked risks and cascading effects. *International Journal of Disaster Risk Reduction*, 30, 281–291. doi:10.1016/j.ijdrr.2018.01.029

Kaminski, I., 21 December 2019. *Dutch supreme court upholds landmark ruling demanding climate action*. Guardian Australia. Available from: https://www.theguardian.com/world/2019/dec/20/dutch-supreme-court-upholds-landmark-ruling-demanding-climate-action [Accessed 16 December 2020].

Kelman, I., Gaillard, J.C., and Mercer, J., 2015. Climate change's role in disaster risk reduction's future: beyond vulnerability and resilience. *International Journal of Disaster Risk Science*, 6, 21–27. doi:10.1007/s13753-015-0038-5

Kim, S.-Y. and Thurbon, E., 2015. Developmental environmentalism: explaining South Korea's ambitious pursuit of green growth. *Politics & Society*, 43 (2), 213–240. doi:10.1177/0032329215571287

Kirkpatrick, D., Apuzzo, M., and Gebrekidan, S., 2020. Europe said it was pandemic ready. Pride was its downfall. *New York Times*, 20 July. Available from: https://www.nytimes.com/2020/07/20/world/europe/coronavirus-mistakes-france-uk-italy.html [Accessed 31 July 2020].

Krasner, S., 1999. *Sovereignty: organized hypocrisy*. Princeton: Princeton University Press.

Lawrence, J., 2018. Manufacturing biopolitical disaster: instrumental (ir)rationality and the deepwater horizon disaster. *In*: J. Lawrence and S. Weibe, eds. *Biopolitical disaster*. London and New York: Routledge, 15–29.

Lenton, T., *et al.*, 2019. Climate tipping points – too risky to bet against Comment. *Nature*, 575, 592–595. doi:10.1038/d41586-019-03595-0

Lin, T., 2015. Governing natural disasters: state capacity, democracy, and human vulnerability. *Social Forces*, 93 (3), 1267–1300. doi:10.1093/sf/sou104

Litfin, K., ed., 1998. *The greening of the Sovereignty in world politics*. Cambridge: MIT Press.

Meadowcroft, J., 2005. From welfare state to ecostate. *In*: J. Barry and R. Eckersley, eds. *The state and the global ecological crisis*. Cambridge, MA: MIT Press, 3–23.

Millar, J. and Pescaroli, G., 2018. Psychosocial capacity building in response to cascading disasters: a culturally informed approach. *International Journal of Disaster Risk Reduction*, 30, 164–171. doi:10.1016/j.ijdrr.2018.04.018

NASA, 2019. *Artic fires fills the sky with soot*. Earth Observatory. US, NASA. Available from: https://earthobservatory.nasa.gov/images/145380/arctic-fires-fill-the-skies-with-soot [Accessed 25 June 2020].

O'Keefe, P., Westgate, K., and Wisner, B., 1976. Taking the naturalness out of natural disasters. *Nature*, 260 (5552), 566–567. doi:10.1038/260566a0

Okereke, C., 2008. Equity norms in global environmental governance. *Global Environmental Politics*, 8 (3), 25–50. doi:10.1162/glep.2008.8.3.25

Pantuliano, S., 2020. Editorial: reflecting on a decade at the helm of *disasters*. *Disasters*, 45 (1), 3–4. doi:10.1111/disa.12469

Park, S. and Kramarz, T., eds., 2019. *Global environmental governance and the accountability trap*. Cambridge, MA: MIT Press.

Parrado, S., 2018. The culture of risk regulation: responses to environmental disasters. *Regulation and Governance*. doi:10.1111/rego.12214

Pauly, L. and Grande, E., 2007. Reconstituting political authority. *In*: E. Grande and L. Pauly, eds. *Complex Sovereignty*. Toronto: University of Toronto Press, 3–21.

Pennisi Di Floristella, A., 2016. Dealing with natural disasters. *The Pacific Review*, 29 (2), 283–305. doi:10.1080/09512748.2015.1013498

Pescaroli, G., 2018. Perceptions of cascading risk and interconnected failures in emergency planning: implications for operational resilience and policy making. *International Journal of Disaster Risk Reduction*, 30, 269–280. doi:10.1016/j.ijdrr.2018.01.019

Readfearn, G., 17 November 2019. *'What could I have done?' The scientist who predicted the bushfire emergency four decades ago*. The Guardian Australia. Available from: https://www.theguardian.com/australia-news/2019/nov/17/what-could-i-have-done-the-scientist-who-predicted-the-bushfire-emergency-four-decades-ago [Accessed 25 June 2020].

Ritchie, H., 2014. *Natural disasters*. OurWorldInData.org, University of Oxford. Available from: https://ourworldindata.org/natural-disasters [Accessed 25 June 2020].

Schrijver, N., 2010. *Development with destruction: the UN and global resource management*. Bloomington: Indiana University Press.

Scott, J., 1998. *Seeing like a state: how certain scheme to improve the human condition have failed*. New Haven and New York: Yale University Press.

Shipper, L. and Pelling, M., 2006. Disaster risk, climate change and international development: scope for, and challenges to, integration. *Disasters*, 30 (1), 19–38. doi:10.1111/j.1467-9523.2006.00304.x

Smith, M., 2009. Against ecological Sovereignty: agamben, politics and globalisation. *Environmental Politics*, 18 (1), 99–116. doi:10.1080/09644010802624843

Steffen, W., *et al.*, 2015. Planetary boundaries: guiding human development on a changing planet. *Science*, 347 (6223), 1259855. doi:10.1126/science.1259855
UNDRR, 2018. *Economic losses, poverty and disasters.* Geneva, Switzerland: United Nations Office for Disaster Risk Reduction (UNDRR).
UNDRR, 2019. *Global assessment report on disaster risk reduction.* Geneva, Switzerland: United Nations Office for Disaster Risk Reduction (UNDRR).
UNISDR, 2015. *Sendai framework for disaster risk reduction, 2015-2030.* Geneva, Switzerland: United Nations International Strategy for Disaster Reduction.
UNISDR, 2017. *Terminology.* Geneva, Switzerland: United Nations International Strategy for Disaster Reduction.
United Nations, 15 May 2020. *Supply chain and COVID-19: UN rushes to move vital equipment to frontlines.* Department of Global Communications. United Nations, Geneva. Available from: https://www.un.org/en/coronavirus/supply-chain-and-covid-19-un-rushes-move-vital-equipment-frontlines
Valdivieso, P. and Andersson, K., 2017. Local politics of environmental disaster risk management: institutional analysis and lessons from Chile. *Journal of Environment & Development*, 26 (1), 51–81. doi:10.1177/1070496516685369
Warner, R., 2005. *Discourses of global environmental governance from colonialism to the 21st century.* Canada: York University.
Wendt, A., 1992. Anarchy is what states make of it: the social construction of power politics. *International Organization*, 46 (2), 391–425. doi:10.1017/S0020818300027764
Wendt, A., 1999. *Social theory of international politics.* Cambridge: Cambridge University Press.
Widmaier, W., 2016. *Economic ideas in political time: the rise and fall of economic orders from the progressive era to the global financial crisis.* Cambridge: Cambridge University Press.
Widmaier, W., Blyth, M., and Seabrooke, L., 2007. Exogenous shocks or endogenous constructions? The meanings of wars and crises. *International Studies Quarterly*, 51, 747–759. doi:10.1111/j.1468-2478.2007.00474.x
Wisner, B., 2001. Disasters: what the United Nations and its world can do. *Global Environmental Change Part B: Environmental Hazards*, 3 (3), 125–127. doi:10.1016/S1464-2867(02)00005-0
World Commission on Environment and Development, 1987. *Our common future.* Geneva: United Nations.
World Meteorological Organization, 2020, *WMO statement on the state of the global climate in 2019.* Geneva, Switzerland: World Meteorological Organization, WMO No – 1248.
World Meteorological Organization, 2 December 2020b. *2020 on track to be one of three warmest years on record.* WMO Press Release, number 02122020. Available from: https://public.wmo.int/en/media/press-release/2020-track-be-one-of-three-warmest-years-record [Accessed 14 December 2020].
Young, O., *et al.*, 2006. The globalisation of socio-ecological systems: an agenda for scientific research. *Global Environmental Change*, 16, 304–316. doi:10.1016/j.gloenvcha.2006.03.004
Zelli, F. and Van Asselt, H., 2013. The institutional fragmentation of global environmental governance. *Global Environmental Politics*, 13 (3), 1–13. doi:10.1162/GLEP_a_00180

Planetary disasters: moving the UN disaster risk reduction framework into cosmopolitised reality

Sabine Selchow

ABSTRACT
I aim to advance the governance of disasters by making it correspond with cosmopolitised reality. Following Ulrich Beck, this correspondence is essential for handling disasters in a constructive way because cosmopolitised reality does not simply challenge existing institutional settings; it reflexively undermines them. I achieve my aim by intervening into a distinct part of the disaster governance discourse, the UN Disaster Risk Reduction Framework discourse. Grounded in a systematic analysis of this discourse, I identify the concept 'systemic risk' as a discursive hook to introduce a new concept, 'planetary disaster'. 'Planetary disaster' inscribes the distinctly integrated nature of cosmopolitised reality into the idea of 'disaster'. With that, it opens new horizons in imagining disaster governance and its institutions.

Introduction

My aim is to advance the governance of disasters so that it corresponds with cosmopolitised reality. Following Ulrich Beck (2004, 2009), this aim is motivated by the understanding that cosmopolitised reality is distinctly integrated and holds an uncertainty that does not simply *challenge* existing institutional settings but reflexively *undermines* them. Hence, without integrating this reality into our efforts to govern disasters we will keep reproducing a social reality in which our institutional tools are inherently inadequate and, indeed, are undermined by themselves and the social reality they produce.

One way to advance disaster governance is to intervene into the discourse that constitutes it. With 'discourse' I mean the web of meanings through which social reality – such as the institutions that shape disaster governance – is imagined and comes into being.[1]

Understood in this way, the disaster governance discourse is extensive and not easily to be grasped. Hence, to intervene into it in a way that is productive and concrete, I focus on a distinct part of it: the UN Disaster Risk Reduction (DRR) Framework discourse. This is the web of meanings that is produced under the auspices of the UN Office for Disaster Risk Reduction. I chose this part of the disaster governance discourse because the UN DRR Framework discourse is influential and flexible. As I will show, flexible means that it is relatively open for discursive interventions.

To productively intervene into a discourse, does not simply mean to critique the existing web of meanings and the world this web allows social actors to imagine. It requires identifying existing openings and widening them by weaving into the existing web alternative definitions and concepts.

Hence, at the heart of my effort is the systematic study of how the UN DRR Framework discourse handles cosmopolitised reality. The aim and value of this is to identify if there are features that could be built on and extended in an effort to reimagine institutions so that they correspond with cosmopolitised reality. For this empirical endeavour I operationalise Beck's theoretical work for empirical purposes and apply it to the *Global Assessment Report on Disaster Risk Reduction 2019*. This is a 160,000-word-document that I conceptualise as the materialisation of the current UN DRR Framework discourse.

My analysis brings to light three promising features in the current discourse: the concepts 'people-centred data', 'contexts', and 'systemic risk'. I put forward these three identified features as the foundation for future interventions into the disaster governance discourse. I take up one of these features, the concept 'systemic risk' as it is used in the UNDRR Framework discourse. I identify it as a promising opening, a hook to weave into the UN DRR Framework discourse a new concept, which I call 'planetary disaster'. 'Planetary disaster' enables us to imagine realities in addition to and beyond what the system-metaphor at the heart of 'systemic risk' allows us to imagine. It forces a profound breakdown of guiding dichotomies, hence, inscribes the distinctly integrated nature of cosmopolitised reality into the idea of 'disaster'. As such, 'planetary disaster' opens new horizons in imagining disaster governance and its institutions.

I start by introducing the UN DRR Framework discourse, and explain what I mean by 'cosmopolitised reality'. For the latter, I build on the theories of reflexive modernisation and cosmopolitisation, developed by Ulrich Beck (2004), Beck (2009)). While Beck's theory of reflexive modernisation is sporadically used, if not in debates on disaster risk reduction narrowly then in environmental discourses broadly,[2] the concept 'cosmopolitisation' has yet to enter the debate. I aim to introduce it.

The UN disaster risk reduction framework discourse

The UN DRR Framework discourse is the web of meanings that is produced under the auspices of the UN Office for Disaster Risk Reduction. It is an extensive web of meanings. Countless international, regional, national and local actors are involved in its production and reproduction, through the introduction of new concepts, definitions and frames. And it is a robust web of meanings. This is because it is grounded in regular UN World Conferences and their associated framework documents. To date, there have been three conferences, in 1994, 2005 and 2015, which brought out three framework documents, the *Yokohama Strategy and Plan of Action for a Safer World* (Yokohama 1994), the *Hyogo Framework for Action 2005–2015* (Hyogo 2005), and the *Sendai Framework for Disaster Risk Reduction 2015–2030* (Sendai 2015). But, of course, the web of meanings that I call 'UN DRR Framework discourse' comes into being through more than these documents. It is the numerous papers, reports and other symbolic interventions, produced by academics, civil society actors and political decision makers that make up this web of meanings. The Framework discourse's robustness is reinforced by numerous institutional nodes, such as the biennial multi-stakeholder forums (UN Global Platforms for Disaster Reduction), which have taken place since 2007, and numerous Ministerial Conferences and Regional Platforms, as well as regular publications, like the *Words into Action* series and, prominently, the biennial *Global Assessment Report on Disaster Risk Reduction* (GAR). The UNDRR-managed knowledge sharing platform *PreventionWeb*(URL) (2020), launched in 2007, has become an authoritative global source of knowledge on how to understand and govern disaster risk. In short, the UN DRR Framework discourse is a web of meanings that is produced by many different social actors in many different, heavily institutionalised, contexts and through a sophisticated knowledge pooling and production process, supported by the *PreventionWeb*-platform. Above and beyond, the UN DRR Framework discourse is framed to sit at the heart of the 2030 Agenda of sustainable development. One of its textual manifestations, the 2015 Sendai Framework document, is presented as nothing less than 'the connecting tissue for the 2030 Agenda for Sustainable Development, the Paris Agreement, the New Urban Agenda, the Addis Ababa Action Agenda and the Agenda for Humanity' (GAR 2019: iii).

This heavily institutional nature makes the UN DRR Framework discourse, the web of meanings produced under the auspices of the UN Office for Disaster Risk Reduction, an influential aspect of the broader disaster governance discourse. As such, it is a fruitful focus for my purposes; intervening into it reaches relatively far.

And there is a second feature of the UN DRR Framework discourse that makes it productive for the discursive intervention that I have in mind for pushing disaster governance towards cosmopolitised reality. This is that it is a relatively flexible discourse. The UN DRR Framework discourse is flexible in the sense that it is expressly open to meaningful change.

Implied in the premise that discourses, webs of meanings, are a constitutive aspect of social reality is the widely accepted fact that 'social reality does not fall from heaven but that human agents construct and reproduce it through their daily practices' (Risse 2007, p. 128). While reality exists outside semiotic signs, it is not accessible to social actors without having been made meaningful. Hence, making the world meaningful *is* making the world. The process of world making is complex, messy and shaped by power. This is because, rather than being referential, meanings are conventional. Meaning is a 'constant flickering of presence and absence together' (Eagleton 1983, p. 128), filtering through language like a web-like shadow (also Derrida 1981). And yet, despite their principal arbitrariness, meanings and language are relatively stable. We *do* communicate and, of course, we *do* 'know' what something 'means'. This is because conventions and rules restrict which shadow of a meaning is to be associated with which signifier. While, in principle, everything is sayable, thinkable and doable, in practice, social ratification and rejection tame and restrict it. Foucault (1981, p. 59–60) illustrates this by reference to Gregor Mendel's theory of hereditary traits. Foucault (ibid. 60) wonders 'how the botanists or biologists of the nineteenth century managed not to see that what Mendel was saying was true' and concludes that while Mendel 'spoke the truth, [...] he was not "within the true" of the biological discourse of his time' (ibid. 61).

> It is always possible that one might speak the truth in the space of a wild exteriority, but one is 'in the true' only by obeying the rules of a discursive 'policing' which one has to reactivate in each of one's discourses. (ibid.)

The UN DRR Framework discourse is such an ordering system that shapes what is 'in the true', for instance by providing definitions of key terms.[3] Indeed, as suggested above, given its heavily institutionalised nature, it is a particularly influential ordering system. And, yet, if we look at it, we see that the Framework discourse is remarkably *flexible* and *open* to change. This is an empirical observation that is grounded in the striking degree of meta-linguistic reflections that feature within the discourse. Express debates about the meanings of key words (prominently 'disaster') are but one example of this reflectiveness (e.g. Kehlman 2018). They demonstrate its flexibility and openness. It is this flexibility and openness which makes the UN DRR Framework discourse a particularly fruitful focus of attention if we want to advance disaster governance; given its flexible constitution it is relatively easy to intervene.

Cosmopolitised reality

Advancing disaster governance, I argue, requires embedding it into fundamentally different parameters. These different parameters need to mirror that contemporary reality – which I understand with Beck (2004), Beck (2009)) as a 'cosmopolitised reality' – is distinctly integrated and holds a distinct uncertainty. This uncertainty does not simply challenge existing institutional settings but reflexively undermines them.

The term 'cosmopolitised reality' is grounded in the social diagnosis developed by Ulrich Beck (1999), Beck (2004), 2009). At the heart of this diagnosis are three theorems, risk society, reflexive modernisation and individualisation, which feed into Beck's theory of 'cosmopolitisation' (see Selchow 2016). Drawing on ideas from across Beck's work, I use the term 'cosmopolitised reality' to describe contemporary reality by highlighting two of its central features: it is distinctly integrated and distinctly uncertain.[4]

Cosmopolitised reality is a distinctly integrated reality because of the process of 'cosmopolitisation' and because of 'global risks' (Beck 2004, 2009). Cosmopolitisation is a social process in which lived realities, cultures, horizons of experience and horizons of expectations are enmeshed, irrespective of national borders. Crucial to understanding the process of cosmopolitisation is that it unfolds beyond and independently of conscious commitments to any form of *normative* cosmopolitanism (Beck 2004).[5] Indeed, cosmopolitisation *happens* in and through daily lives. It is a process that transforms national societies *from within* as it unfolds as a *side-effect* of actions that are targeted at other ends (Beck 2004: 113–114, Beck and Sznaider 2006, p. 9); cosmopolitisation unfolds 'accidentally' and, importantly, for instance, including through actions that are motivated by dividing sentiments, such as the recent evocations of exclusive nationalism, but rely on global social media networks (Beck 2004; also Selchow 2018). As Beck (2004, p. 113) explains, its side effect-nature makes the process of cosmopolitisation 'long-term' and 'irreversible'.

Furthermore, cosmopolitised reality is a distinctly integrated reality because of 'global risks'.[6] To remind us, 'risk' is a modern technology to tame uncertainty. It is a strategy to 'colonise the future' (Giddens 1991: 11) in order to enable action in the present. Risk is a child of modernity in that it is about agency, decisions and, inevitably, responsibility. It only exists in a world in which the future is open to be shaped by human action in the present.

But, of course, the future can never be known – even if it is approached through the logic of risk and modelled through ever more sophisticated risk assessments, based on ever more extensive databases and fine-grained risk-categories, such as those used in the UN DRR Framework discourse (GAR 2019). Risk is always only a way of *imagining* the future and *potential* consequences of present decisions. It is 'non-existent, constructed and fictitious' (Beck 1999, p. 214).

Having sketched the general idea of risk, the term '*global* risk' refers to the potential consequences of *specific* decisions, namely, what Beck (2014, p. 98) calls, 'industrial, that is, techno-economic decisions and considerations of utility'. These are decisions grounded in modern institutions (e.g. nation-state, family) and basic modern principles (e.g. freedom, market dependence, rationality, progress, statehood) (ibid.). For instance, the decision to publicly support the expansion of the German automobile sector to secure national economic growth is an example of such 'industrial, that is, techno-economic decisions and considerations of utility'. Understood in this sense, global risks, that is, the potential consequences of these specific decisions, are integrating because they are potentially 'socially delimited in space and time' (Beck and Grande 2010, p. 418). That they need to be considered as potentially 'socially delimited in space and time' is an empirical observation that picks up innumerable instances of *actual* consequences of past decisions (i.e. not 'risks' but 'hazards' and 'disasters'), such as manifestations of climate change as the (not anymore potential but *actual*) unintended consequences of all those past decisions that fostered CO2-intensive developments, like an expanded automobile sector. Consequently, the crucial point about global risk is that if we were to make 'industrial, that is, techno-economic decisions and considerations of utility' by considering global risk, we would inevitably undermine some of the modern premises on which the technology of risk (and these decisions) are based, especially and importantly the idea that decisions could be made based on the imagination of bounded societies (Beck 1986).

And, yet, not considering global risks means ignoring two defining features of reality. Hence, cosmopolitised reality and its global risks do not just sit square to the modern national container thinking and to a time-bound view of impact and responsibility, they also *undermine* them and other modern principles and concepts. In particular, they sit square and undermine the notion and practices of externalisation that shape much of our modern, national existences. With 'externalisation' I refer to the 'outsourcing' of 'bads' of modern nation state-based lifestyles to other societies and 'global others', such as life threatening working conditions and child labour in the garment and shrimp industries, the export of electronic and plastic waste or, in fact, the externalization of agrarian production to another part of the world, with all its ecological and economic consequences (Lessenich 2016, Brand and Wissen 2018). Cosmopolitised reality and its global risks sit square to all of this because global risks bring 'global others' and 'future others' into the midst of the decision of other 'global others' and 'present others'. This is a reality that unfolds whether or not those involved notice and acknowledge it. To reiterate, what makes this dynamic significant and powerful is that it is not the product of conscious cosmopolitan, normative considerations but the unintended product of the reality of global risks. Consequently, cosmopolitised reality is a reality of a borderless necessity to cooperate (*Kooperationszwang*) as well as interrelation of responsibility (*Verantwortungszusammenhang*) (Beck and Grande 2010, p. 417) – this is not a question of 'should' but a matter of 'is'.

In addition to being distinctly integrated, cosmopolitised reality is distinctly uncertain; this picks up the issue of 'undermining' that I mention above. There are two aspects to this point.

First, given the technological advancements entailed in industrialization and the progress of modernization and modern knowledge production, the potential consequences of 'industrial, that is, techno-economic decisions and considerations of utility', i.e. global risk, are not only to be considered as potentially 'socially delimited in space and time' – the point I made above – but also as potentially falling into the realm of non-knowledge (*Nichtwissen*) (Beck 2009). In short, potentially, knowledge does not only produce things that we do not know *yet* but things that we do not know we do not know and that we might never be able to know (see Wehling 2006).

Second, cosmopolitised reality is distinctly uncertain because there is something reflexive about the process of modernisation. As Beck (1994), Giddens (1990) and other sociologists explain, there is something reflexive about modernisation because in the course of modernisation and the unfolding of modernisation's side-effects (such as climate change) the basic principles of modernity as well as modern basic institutions are 'radicalised' (Beck 1994, 2009). This means that in cosmopolitised reality modern institutions and principles are confronted with the consequences of the *success* of modernisation itself, which, in the course of its successful unfolding, reflexively overturns its own foundations, i.e. its institutions and principles. For example, climate change as the unintended consequence of the *successful* unfolding of modernisation, including its ever advancing knowledge production that made possible, for instance, commercial air travel, undermines its modern foundations, not only by undermining development, as we see it in the context of disasters, but also, more fundamentally, by making the idea of bordered societies, a central modern principle, obsolete. And it does this not as an 'undesired side-effect' that could be tackled by additional knowledge production but as something *inevitably entailed* in modernisation's *success* and its own radicalised institutions, principles and concepts. What is required, then, is to fundamentally rethink how we organise social life and through which concepts we imagine it because it is modern institutions themselves, as a fruit of their *success*, that produce a world that these institutions then cannot deal with. In Beck's (1991: 41) words, modernisation has turned into 'a sub-political "revolutionary system" without a revolutionary program or goal'. As a consequence, lived reality – the horizons of experience and horizons of expectations of 'side-effect cosmopolitisation' – correlates neither with the institutions of modern (national) societies nor with their principles, making contemporary cosmopolitised reality distinctly, one could say, inherently uncertain.

In sum, contemporary reality, cosmopolitised reality, is distinctly integrated and distinctly uncertain. Both, its integrated nature and uncertainty are the *product* of modern institutions, principles and concepts and *challenge* these modern institutions, principles and concepts fundamentally.

A double challenge at the right time

Reimagining the foundations of our modern institutional setting is a major challenge; indeed, it is a double challenge. It is not only about actually doing it but about doing it without having an adequate conceptual toolkit to do so. Imagining institutions, principles and concepts that correspond with cosmopolitised reality cannot be done with the existing conceptual toolkit that comes out of and corresponds with modern reality. Hence, at the heart of the task of rethinking our institutional landscape is the challenge of finding a language that enables us to do so – which needs to be a language that springs off cosmopolitised reality, is shaped by and mirrors its parameters.

To grasp this problematique, Beck (2009; also Beck and Sznaider 2006) distinguishes between what he calls the 'national outlook' and 'methodological nationalism', on the one side, and the 'cosmopolitan outlook' and 'methodological cosmopolitanism', on the other side. 'National outlook', which is held by political actors, and 'methodological nationalism', which is held by scholars, refer to the modern imagination and its conceptual language, at the centre of which is, as I discussed above, the notions of timely and spatial boundedness, especially the boundedness of societies. As I outlined above, it is this language that produces cosmopolitised reality and, simultaneously, stands in the way of imagining institutions that correspond with cosmopolitised reality; it does this while being relatively robust because, as Beck and Sznaider (2006, p. 4) observe, it 'adopts categories of practice as categories of analysis', inscribing these (then seemingly) natural categories further into the making of the social world. Given the inherent problematique of the national outlook and, indeed, its grammar (Beck 2004, p. 133), what is needed instead is an outlook that corresponds with cosmopolitised reality, a 'cosmopolitised outlook'.[7] This outlook entails not just grasping fully contemporary, cosmopolitised reality but imagining adequate institutions, principles and concepts for and from within this reality.

The double challenge entailed in this is that such a move necessitates that we are able to grasp the nuances of cosmopolitised reality, for which we need a language grounded in such a grasp of the nuances of cosmopolitised reality, i.e. grounded in a 'cosmopolitised outlook' – the challenge comes full circle. And yet, it is simply not an option not to tackle the modern problematique; it is an imperative, including and especially for when it comes to the governance of disasters.

The good news is that now is the right time to approach this challenge. Understandings of the cosmopolitised nature of reality and the profound challenges this poses are not new. Beck's seminal study *Risikogesellschaft* was published in 1986. And another 17 years earlier, McHale (1969: 11) argued,

> [t]he models of human society, of our institutions, and of our social capabilities and relatedness, with which we still operate, restrict much of our futures thinking within obsolete historical conditions. We can produce material goods far beyond immediate necessity and have elaborated our organizational capacities beyond those necessary for mere group survival and security, but our outdated models still generate the same stereotyped "dilemma" responses to most of our social problems. The constraining myths that bind us to obsolete forms, old fears, and insecurities may be our most dangerous deterrents. Our traditional attitudes and ideologies are inadequate guides to the future, serving mainly to perpetuate old inequities and insecurities.

As I outlined earlier, the acceptance of ideas does not depend on if they are 'the truth' but if they are 'in the true' of the respective discourse that polices them. Regarding the challenge of tackling the problematique of reflexive modernisation, we are at an important moment in time. There is a discursive opening for moving the notion of 'cosmopolitised reality' and the necessity to rethink our modern institutional environment into 'the true'. There is a growing sense that we need radical new concepts for and of the world. This is apparent, for instance, in the mushrooming literature that questions the modern premises underlying the monetary measure GDP and that query the world this measure brings out (e.g. Jackson 2016, Raworth 2017); it is also apparent in the national security discourse in which the European notion of 'human security' is starting to take hold, undermining the discourse-determining conception of 'inside/outside' (e.g. Kaldor *et al.* 2018); and it is apparent in the UN DRR Framework discourse itself which, as I proposed earlier, is remarkably flexible and open for conceptual changes and discursive innovations.

So how does the UN DRR Framework discourse handle cosmopolitised reality and where, if at all, are features in this discourse that could be extended in an effort to imagine institutions that correspond with cosmopolitised reality?

Analytical strategy

I investigated the UN DRR Framework discourse through an analysis of the *Global Assessment Report on Disaster Risk Reduction 2019* (GAR 2019) and the *Global Assessment Report on Disaster Risk Reduction 2019 Distilled* (GAR 2019 Distilled). It is an analytical choice to take these two documents as materialisations of the current UN DRR Framework discourse. GAR 2019 (158,598 words) is the most recent edition in the *Global Assessment Report* series. The aim of these Reports is to capture and analyse disaster risk patterns, trends and consequences, as well as progress and developments in disaster risk responses. They do this by drawing on a variety of different sources, including tailored case studies and scientific background papers, as well as a DRR database that national agencies deliver in accordance with the UN DRR Framework agreements.

I compiled and used two text corpuses, one with each of the five existing GARs, and one with a compilation of GARs 2009–2013 (204,187 words) and GARs 2015–2019 (241,016 words).[8] The purpose of the second corpus was to ensure that my automated reading captured similarities but also possible differences between the documents that were produced with view to the Hyogo Framework (GARs 2009–2013) and the Sendai Conference and Sendai (2015) (GARs 2015–2019). Using all documents in the GAR series as a reference corpus allowed me to get a sense of the common language in the UN DRR Framework discourse. This enabled me to detect the particularities of the contemporary state of the discourse, as well as those changes over time that were relevant for understanding if and how the UN DRR Framework discourse helps us to push forward the important task of new ways of thinking.

The above sketched double challenge to imagine institutions, principles and concepts that correspond with cosmopolitised reality and go beyond the 'national outlook' does counters a 'traditional' empirical approach that utilises a preset coding scheme. This is because such an approach would reproduce the 'national outlook'. At the same time, as sketched above, we do not yet have the right language and conceptual toolkit that corresponds with cosmopolitised reality and enables us to fully explore this reality, in a way that goes beyond identifying its two main features that I have already outlined. It is in the nature of the challenge that such language cannot solely be developed at the whiteboard. Theory-driven innovations are part of the effort but need to be embedded in insights from within empirical reality, the latter generated as a long-term collective effort through empirical studies that are as inductive as possible (Selchow 2020). And yet, especially the first steps in this endeavour to find and develop a positive cosmopolitised language, do not work without being anchored in the modern world; at least initially, it is not possible to avoid using modern institutions, principles and concepts as the negative in order to identify how cosmopolitised reality is handled in a particular discursive instance, like the UN DRR Framework discourse.[9]

Given this profound epistemological challenge, I developed and applied a dynamic coding scheme. This means that my coding scheme evolved in the process of my analysis, starting with an initial, deductively developed layer and extending it inductively. I combined close and distant reading strategies, assisted by a computer.[10] I used four basic questions to serve as the foundation of this scheme. Addressing the rhetorical dimension of the documents, my questions were: How self-reflective are the texts regarding their role in the production of certainty? How open are they, i.e. where and how do they leave space for other knowledges to enter the production of certainty? Addressing the web of meanings, my questions were: What possibilities does the world hold for rethinking and going beyond modern principles and concepts? Are there alternative ideas to modern principles and concepts in general, and to the modern notions of boundedness, as well as, certainty?

I took two steps. First, I grasped the nature of the world of the UN DRR Framework discourse. Second, I examined this world in view to its cosmopolitised potentials.

The world of the UN DRR Framework discourse

The world of the UN DRR Framework discourse is a world which is oriented towards the tackling of risks. Disaster governance is risk governance. Risks are to be 'prevented', 'curbed' and 'reduced'. Indeed, it is the '*creation* of new risks' that is to be prevented, curbed and reduced.

The motivation to prevent risks is existential. It is existential in a narrow sense for small island development states, as distinct actors in the world of the UN DRR Framework discourse. But it is also existential for 'us', as 'humans':

> Our flexibility must be as dynamic as the changes we hope to survive. (GAR 2019: xi)

> The very survival of humans on the planet is at stake. (GAR Distilled 2019: NB)

The tackling of risks is something that concerns 'us all'; there is a 'we' and an 'our' in the world of the contemporary Framework discourse. The world of the contemporary Framework discourse requires 'collective action', 'collective responsibility' and 'collective efforts' (Figure 1). And there is 'humanity', an entity that did not exist in past worlds of the UN DRR Framework discourse.

Yet, actual risk reduction lays with individuals, especially those in disaster-affected contexts. It is the behaviour and choices of individuals, including the norms which inform behaviour and choices, that need to be changed in an effort to reduce risk. Of course, individuals are not the only actors in the world of the contemporary Framework discourse. Indeed, they stand in direct line with a number of other actors, such as 'communities', 'populations', ' societies', 'communities and governments', 'communities and nations', 'communities and society', 'communities, countries and international actors', 'households and communities', 'households, communities, cities, countries, economies or

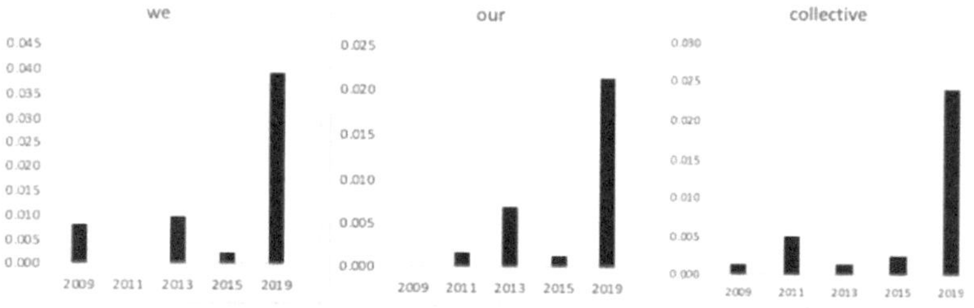

Figure 1. The number of uses of the terms 'we', 'our' and 'collective', relative to the total number of words in each document.

ecologies', 'communities, health systems and economies'. 'Nation-states' do not exist in the world of the contemporary Framework discourse, neither have they existed in the world of past discourses. And neither are NGOs and 'civil society' now accredited with an outstanding role, in comparison to the world of past instances of the Framework discourse (Figure **). What plays an outstanding role, though, is 'governments'. While it is first and foremost individuals who need to adapt their behaviour and choices, it is the task of

> governments to find solutions and influence human behaviour, to successfully prevent the creation and propagation of risk, as well as to rebound from disasters. (GAR Distilled 2019: NB)

More broadly,

> Governments have a responsibility for creating an environment in which people prosper and the planet thrives. (GAR Distilled 2019: NB)

The world of the contemporary Framework discourse is highly integrated. It is 'connected', 'interconnected', 'hyperconnected', 'interdependent', 'interrelated', shaped by 'feedback loops', 'reverberation' and 'cascading' dynamics (see Figure 2), and stretching beyond jurisdictions (GAR 2019, p. 61).

In particular, it is integrated due to the nature of risk and the fact that risk, having replaced disaster, is inscribed in development:

> As presented in previous GARs and reiterated in this edition, it is time to cast off the notion that risk is exogenous to development, something that can be reduced simply by complementing development with risk reduction measures. (GAR 2019, p. 333)

In addition to being integrated, the world of the contemporary Framework discourse is complex and multi-layered (Figure 3); it is full of 'multiple dimensions', 'multiple scales', 'multiple impacts', 'multiple challenges', 'multiple levels', 'multiple hazards'.

The world is ordered across different categories, with global, international, regional, national, subnational and local 'levels' being the dominant ordering

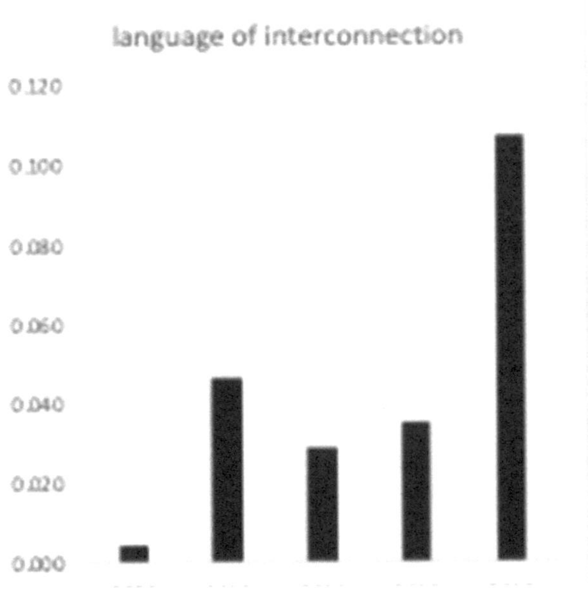

Figure 2. The number of uses of terms associated with the idea of interconnection, relative to the total number of words in each document.

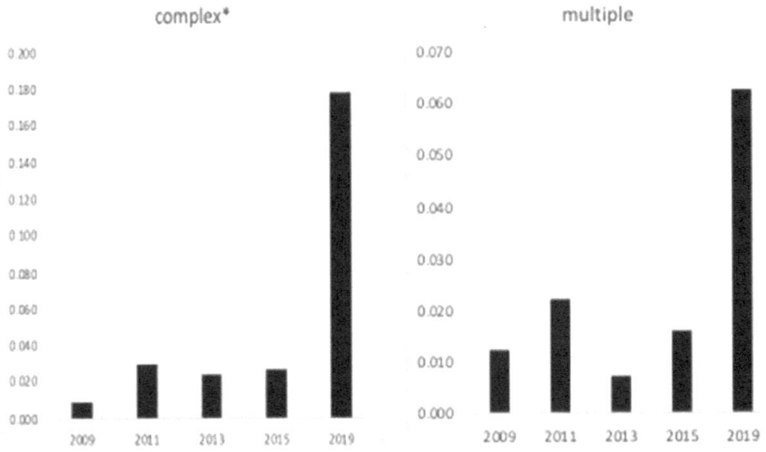

Figure 3. The number of uses of the 'complex*' and 'multiple', relative to the total number of words in each document.

systems. But the world of the contemporary Framework discourse also consists of households as a central reference point for risk data collection. Furthermore, it consists of three foundational components. These are 'systems', 'networks', and 'contexts', such as 'local contexts', 'country contexts' or 'the context of rapid climatic change' (GAR 2019, p. 48) (Figure **). The contexts are not static but 'rapidly changing' and 'dynamic'. They shape the world and disaster related risks essentially, hence, are an integral part of understanding the governance of these risks.

> None of the "wicked problems", as described by IPCC and multiple other scientific bodies, that are currently pressuring policymakers to try new approaches to meet today's challenges, can be understood with reductionist approaches. In other words, the deliberate simplification of a problem and its causes by removing it from its context renders the understanding and ensuing solution obsolete. The issues with which we are confronted are wrapped in contextual interdependencies that require an entirely different approach in assessment and action. (GAR 2019, p. 60)

The 'systems' are not just components of this world either. They, too, shape approaches to the world's challenges and its very nature, including the nature of risks. Consequently, the risks in the world of the contemporary UN DRR Framework discourse are no longer 'disaster risks' but 'systemic risks' (Figure 4). Both, networks and systems have agency and are personified; they have behaviours (e.g. GAR 2019: 36, p. 46).

Risks are the central challenges in the world of the contemporary Framework discourse. There are hardly 'dangers' in this world. Yet, there are 'threats' and, importantly, there are 'accidents' as well as, Natural Hazard Triggering Technological (NATECH) Disasters, i.e. 'cascading technological side effects of natural hazards' (GAR 2019, p. 126) (Figure 5).

But risk is more than the focus of attention and action. It is actually inscribed in how things are done. There is 'risk-informed', 'risk-based', 'risk-reducing' and 'risk-sensitive' actions.

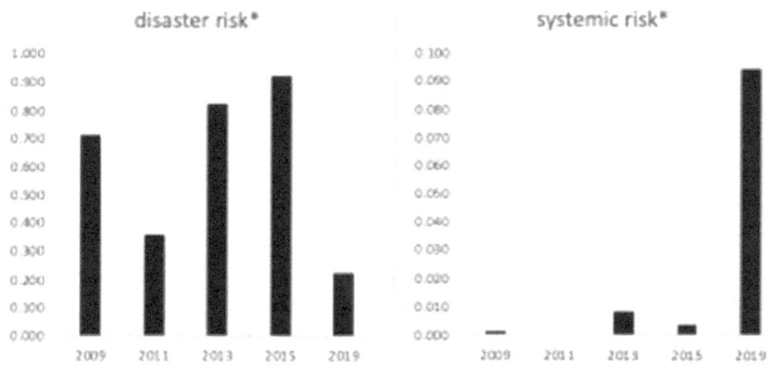

Figure 4. The number of uses of terms 'disaster risk*' and 'systemic risk*', relative to the total number of words in each document.

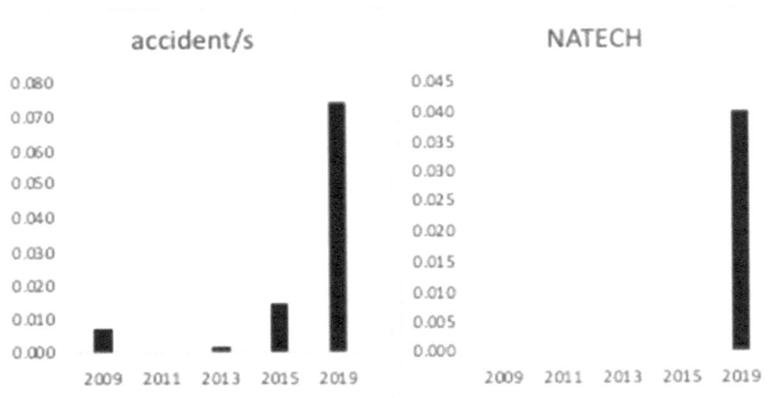

Figure 5. The number of uses of the terms 'accident/s' and 'NATECH', relative to the total number of words in each document.

In the world of the contemporary Framework discourse, the key to tackling systemic risks is data generation, data sharing, open data, good data, contextual data, big data, data that is not 'fragmented, non-universal, incommensurable and biased' (GAR 2019: vi), but rather data that is 'timely, accurate, disaggregated, people-centred and accessible' (ibid.); data that helps to 'understand in finer detail how shocks affect people's lives in a systemic way' (GAR 2019: vii). With the data-driven approach comes an additional actor, an additional 'we'; this 'we' are the data scientists who, first, 'must understand in finer detail how shocks affect people's lives' in order to then 'support governments to find solutions and influence human behaviour.'

Building on appropriately 'good data', a collaborative, interdisciplinary, indeed, 'transdisciplinary, integrated, multisectoral research engaging non-traditional counterparts' (GAR 2019: v), moving beyond 'silos', is needed to address systemic risks. It is an approach that does not only acknowledge and deal with complexity but also addresses the 'certainty of near-term non-linear changes' (GAR 2019, p. 70) and, indeed, is prepared to respond to 'surprises'. 'Surprise is the new normal' in the world of the contemporary UN DRR Framework discourse (GAR 2019: iv), and a profound uncertainty is accepted.

And yet, the world is not completely different-in-kind but at a stage in a linear historical development:

> Our planet, circumstances, needs and choices have always evolved and changed. (GAR Distilled 2019: NB)

Moving the UN DRR framework discourse into cosmopolitised reality

Looking at the world of the UN DRR Framework discourse with a view to how it handles cosmopolitised reality, it is relatively easy to identify that it is deeply embedded in and reproduces modern reality. For instance, there is the notion of historical linearity in the development of the world that places it as a different-in-degree-extension of the past. Although the 'nation-state' does not appear in the world of the Framework discourse it is 'natural', in the sense of unquestioned, that national governments are the ultimate active actors; a 'containered' world is reproduced – also through the traditional categories of international, national, regional etc. And, of course, the world of the Framework discourse turns around *the* modern technology, namely risk; the logic of risk is even inscribed into how things are done. More broadly, the world of the Framework discourse is shaped by an imperturbable belief in the power of knowledge production. To govern disaster adequately requires more of the right data and ever more sophisticated analyses. There is reflection on the limits of the modern taming of uncertainty. Yet, neither 'global risks', the core feature at the heart of cosmopolitised reality, nor the reflexive backfiring of institutions, principles and concepts exist in the world of the UN Framework discourse. The problem of uncertainty, even non-knowledge, is acknowledged but remains an issue to be tackled from within modern reality. The discourse does not question modern premises and does not propel the world into cosmopolitised reality. Overall, disruptive ideas and discursive innovations that might facilitate imagining institutions, principles and concepts that correspond with cosmopolitised reality are not readily apparent in the world of the Framework discourse.

And yet, there are seeds to locate and cultivate. Despite being tamed by the scale of international, national, and local, the kind of interconnectedness that shapes the world of the Framework discourse is relatively profound and sweeps over the containers and inside-outside dichotomies that shape modern reality. Due to the absence of 'global risk', the world of the Framework discourse is not as distinctly integrated as cosmopolitised reality. Yet, it brings out an 'us' and 'humanity' as distinct actors. This holds the possibility of escaping the logic of externalisation that, as discussed above, runs modern reality. Furthermore, the existence of 'accidents' and 'surprises' at least shake up the notion of modern certainty and challenge 'traditional' modern governance principles and premises.

Beyond these seeds, there are three features that are particularly interesting and promising in terms of serving as discursive opening towards cosmopolitised reality.

First, there is the idea of people-centred data. In the world of the Framework discourse it serves as a means to generate better, in the sense of, fine-grained understandings of the lived realities around disasters and disaster risks. What is intriguing about it in regard to a possible move into cosmopolitised reality is that people-centre data counters the idea of national data, as one of the most prominent manifestations of methodological nationalism (Otto *et al.* 2015) and modern reality. In a way, the term 'people-centred' in 'people-centred data' has an effect similar to the one that the term 'human' has in the European notion of 'human security': it provides pathways and material realities beyond the inside/outside dichotomies that shape the modern, international order (e.g. Kaldor *et al.* 2018).

The second promising feature are 'contexts'. In the world of the Framework discourse, contexts bring out distinct lived realities. Contexts are not just place-based but also issue-related. They are not static but themselves moving and dynamic. Hence, because of 'contexts', in the world of the Frameworks discourse social reality is profoundly fluid and flexible. 'Contexts' undermine the idea of static knowledge in a way that goes beyond the modern perception that insights are always subject to revisions. Through dynamic contexts, knowledge production becomes even more, one could say essentially flexible and, indeed, inevitably inclusive – grasping contexts can only be a collaborative endeavour. Above and beyond, contexts entail the possibility to escape the national containered thinking entailed in methodological nationalism. As such, it constitutes a tool to map cosmopolitised reality.

The third promising feature is the notion of 'systemic risk'. As suggested above, with the reliance on 'risk' the world of the Framework discourse relies on a prominent modern technology and inevitably and problematically reproduces modern reality. Yet, the adjective 'systemic' challenges the modern notion of 'risk'. It does this by taking 'risk' out of the national frame it is traditionally embedded in and positions it in an interconnected world of interacting systems. Indeed, given that in the world of the Framework discourse 'systems', such as ecological systems, are actors, i.e. have behaviours, the concept 'systemic risk' brings into 'disaster governance through risk' an entity that can actually not be adequately handled with the risk technology; it both permits and undermines risk. In the same vein, 'surprises' permits but challenge the premises on which the technology of risk is built. Hence, similar to cosmopolitised reality the reality of the world of the Framework discourse is shaped but a lingering sense of uncertainty. In this sense, although deeply embedded in modern reality, the notion of 'systemic risk' also entails the possibility of reflecting on the principles that underlie risk. Yet, while promising in some respects and being profoundly integrated, the systemic world of the Framework discourse is also tamed through its systems, as systems are autonomous entities. Systems interact but do not enmesh.

I argue that these three features are important toeholds to be exploited in an attempt to move the UN DRR Framework discourse into cosmopolitised reality and reimagine modern institutions, principles and concepts. I propose them as subjects for targeted future studies and concerted debate.

Here, I take up one of them, the notion of 'systemic risk' and use it as a discursive hook. Acknowledging the promises but also the problems of 'systemic risk', and acknowledging that it is the guiding concept in the world of the Framework discourse (disaster governance is systemic risk governance in this world), I want to pair 'systemic risk' with a second concept as a double guiding concept for disaster governance. This second concept is 'planetary disaster'. 'Planetary disaster' necessitates additional adjustments to what it means to govern disasters that go beyond what 'systemic risk' implies; it moves the Framework discourse and disaster governance closer to cosmopolitised reality.

I use the adjective 'planetary' in line with a growing number of scholars who turn to 'planetary' in order to avoid 'global' (e.g. Lees, Shin and Lopez-Morales 2016; Friedman 2015; also Rockefeller Foundation–Lancet Commission on planetary health 2015).[11] Here, 'planetary' does not refer to a *scale* or, in a narrow sense, to the *planet*, as in the concept of 'planetary boundaries' (Rockström et al 2009), but as a way to reimagine social and political relations, replacing the adjective 'international', while alluding to the notion of the (materiality of the) planet but also to the ethos of being, suggested by Spivak (2015) under the label 'planetarity'. With 'planetarity' Spivak uses, what she calls, an 'untranslatable' word that is 'untranslatable, because it cannot be referred back to a reference object that symbolically tames it. 'Planetarity' remains 'ungraspable' and makes it possible to imagine a new kind of subject, the 'planetary subject', in the context of which

> alterity remains underived from us; it is not our dialectical negation, it contains us as much as it flings us away — and thus to think of it is already to transgress, for, in spite of our forays into what we render through metaphor, differently, as outer and inner space, it remains that what is above and beyond our own reach is not continuous with us as it is not, indeed, specifically discontinuous. (Spivak 2015, p. 292)

Linking the above to the term 'disaster', the adjective 'planetary' modifies this term in a way that it holds the imperative to imagine the world beyond the 'international', alludes to the planet and, importantly, holds the possibility of a radical dissolution of dichotomies and modern grounds. This includes especially the idea of externalisation.

'Planetary disaster' inscribes into 'disaster', hence, into (planetary) disaster governance, the necessity of flexibility, openness and reflectiveness. At the same time, it captures the understanding of disasters as the most obvious manifestation of the clash between cosmopolitised reality and existing modern institutional settings. 'Planetary disaster' captures the notion that disasters are the product of past 'industrial, that is, techno-economic decisions and considerations of utility' that did not consider global risks; in other words, that disasters are the product of past decisions grounded in and guided by modern principles and concepts, such as the notion of spatial and timely boundedness. This means that the notion 'planetary disaster' necessitates disaster governance to come in at an earlier point in time, namely, to address the modern principles and concepts that guide these decisions to begin with. Governing not just 'systemic risks' but also 'planetary disasters' then constitutes a radical first step for disaster governance in moving towards cosmopolitised reality.

Notes

1. For a discussion of different uses of the term 'discourse' see Selchow (2017).
2. In this journal, see, for instance, Mol (1996), Bäckstrand (2004).
3. For definitions and lexical meanings see Selchow 2017.
4. For the following see Selchow 2016, 2020.
5. This is why the adjective 'cosmopolit-*ised*' is used not 'cosmopolit-*an*'.
6. For the following see Selchow 2020.
7. In order to be coherent and avoid misunderstandings, the adjective 'cosmopolitised' is more adequate than Beck's 'cosmopolitan' (see Selchow 2016).
8. Being a summary of GAR 2019 I did not include GAR 2019 Distilled in my automated readings.
9. For the above and in detail see Selchow 2020.
10. I used *AntConc*.
11. See further Selchow 2017.

Disclosure statement

No potential conflict of interest was reported by the author(s).

References

Bäckstrand, K., 2004. Scientisation vs. Civic expertise in environmental governance: eco-feminist, eco-modern and post-modern responses. *Environmental Politics*, 13 (4), 695–714. doi:10.1080/0964401042000274322

Beck, U., 1986. *Risikogesellschaft: Auf dem Weg in eine andere Moderne*. Frankfurt/ M.: Suhrkamp.

Beck, U., 1991. *Ecological enlightenment: Essays on the politics of risk society*. Amherst, MA: Humanity Books.

Beck, U., 1994. The reinvention of politics. *In*: U. Beck, A. Giddens, and S. Lash, eds. *Reflexive modernization*. Stanford, CA: Stanford University Press, 1–55.

Beck, U., 1999. *World risk society*. Cambridge: Polity.

Beck, U., 2004. *Der kosmopolitische Blick oder: krieg ist Frieden*. Frankfurt am Main: Suhrkamp.

Beck, U., 2009. Cosmopolitanization without cosmopolitans. *In*: K. Ikas and G. Wagner, eds. *Communicating in the third space*. London: Routledge, 11–25.

Beck, U., 2014. *Risk society: towards a new modernity*. London: SAGE.
Beck, U. and Grande, E., 2010. Varieties of second modernity. *The British Journal of Sociology*, 61 (3), 409–443. doi:10.1111/j.1468-4446.2010.01320.x
Beck, U. and Sznaider, N., 2006. Unpacking cosmopolitanism for the social sciences: a research agenda. *British Journal of Sociology*, 57 (1), 1–23. doi:10.1111/j.1468-4446.2006.00091.x
Brand, U. and Wissen, M., 2018. *The limits to capitalist nature*. London: Rowman & Littlefield.
Derrida, J., 1981. *Positions*. London: Athlone.
Eagleton, T., 1983. *Literary theory: an introduction*. Oxford: Blackwell.
Foucault, M., 1981. The order of discourse. *In*: R. Young, eds. *Untying the text: a post-structuralist reader*. London: Routledge, 48–78.
Friedman, S., 2015. *Planetary modernisms: Provocations on modernity across time*. New York: Columbia University Press.
GAR, 2019. *Global assessment report on disaster risk reduction 2019*. UNDRR, Available from: https://gar.unisdr.org [accessed 26 April 2020].
GAR Distilled, 2019. *Global assessment report on disaster risk reduction 2019 distilled*. UNDRR, Available from: https://gar.undrr.org/sites/default/files/gar19distilled.pdf [accessed 26 April 2020].
Giddens, A., 1990. *The consequences of modernity*. Stanford: Stanford University Press.
Giddens, A., 1991. *Modernity and self-identity: Self and society in the late modern age*. Stanford: Stanford University Press.
Hyogo, 2005. *Hyogo framework for action 2005–2015: building the resilience of nations and communities to disasters*. UNDRR, Available from: https://www.undrr.org/publication/hyogo-framework-action-2005-2015-building-resilience-nations-and-communities-disasters [accessed 26 April 2020].
Jackson, T., 2016. *Prosperity without growth*. London: Routledge.
Kaldor, M., Rangelov, I., and Selchow, S., eds., 2018. *EU global strategy and human security: rethinking approaches to conflict*. London: Routledge.
Kehlman, I., 2018. Lost for words amongst disaster risk science vocabulary? *International Journal of Disaster Risk Science*, 9, 281–291. doi:10.1007/s13753-018-0188-3
Lees, L., Shin, H. B. and Lopez-Morales, E., 2016. *Planetary gentrification*. Cambridge: Polity Press.
Lessenich, S., 2016. *Neben uns die Sintflut: die Externalisierungsgesellschaft und ihr Preis*. Berlin: Hanser.
McHale, J., 1969. *The future of the future*. New York: George Braziller.
Mol, A., 1996. Ecological modernisation and institutional reflexivity: environmental reform in the late modern age. *Environmental Politics*, 5 (2), 302–323. doi:10.1080/09644019608414266
Otto, I., Biewald, A., Coumou, D. et al., 2015. Socio-economic data for global environmental change research. *Nature Climate Change*, 5, 503–506. doi:10.1038/nclimate2593
Potter, J., 1997. Discourse analysis as a way of analysing naturally occurring talk. *In*: D. Silverman, eds. *Qualitative research*. London: Sage, 200–222.
PreventionWeb (URL), Available from: https://www.preventionweb.net/sendai-framework/gar [accessed 26 April 2020].
Raworth, K., 2017. *Doughnut economics*. London: Penguin.
Risse, T., 2007. Social constructivism meets globalization. *In*: D. Held and A. McGrew, eds. *Globalization theory: approaches and controversies*. Cambridge: Polity Press, 126–147.

Rockström, J., et al., 2009. Planetary boundaries: exploring the safe operating space for humanity. *Ecology and Society*, 14, 2. Available from: https://www.jstor.org/stable/pdf/26268316.pdf?refreqid=excelsior%3A43f3a011be8cd7be120e0a77bcb5d86b [accessed 30 December 2020].

Selchow, S., 2016. The paths not (Yet) taken: Ulrich beck, the 'cosmopolitised world' and security studies. *Security Dialogue*, 47 (5), 369–385. doi:10.1177/0967010616647859

Selchow, S., 2017. *Negotiations of the 'New World': The omnipresence of global as a political phenomenon*. Bielefeld: Transcript.

Selchow, S., 2018. Breaking the shackles of the national mindset in a polarised world', The Conversation, 30 March 2018, Available from: https://theconversation.com/breaking-the-shackles-of-the-national-mindset-in-a-polarised-world-90009 [accessed 30 December 2020

Selchow, S., 2020. Starting somewhere different: methodological cosmopolitanism and the study of world politics. *Global Networks*, 3, 544–563.doi:10.1111/glob.v20.3

Sendai, 2015. *Sendai framework for disaster risk reduction 2015–2030*, UNDRR, Available from: https://www.undrr.org/publication/sendai-framework-disaster-risk-reduction-2015-2030 [accessed 26 April 2020].

Spivak, G.C., 2015. Planetarity. *Paragraph*, 38 (2), 290–292.

Wehling, P., 2006. *Im Schatten des Wissens? Perspektiven der Soziologie des Nichtwissens*. Konstanz: UVK.

Yokohama, 1994. *Yokohama strategy and plan of action*, UNISDR, Available from: https://www.unisdr.org/files/8241_doc6841contenido1.pdf [accessed 26 April 2020].

OPEN ACCESS

Rehabilitating Ranger uranium mine: scientific uncertainty, deep futures and the production of ignorance

Rebecca Lawrence

ABSTRACT
This research explores contestations surrounding the rehabilitation of the Ranger Uranium Mine in the Northern Territory of Australia. I highlight how particular scientific knowledges are privileged throughout the rehabilitation process, but only so long as the rehabilitation problems at hand are deemed manageable. I also argue that the implications of the immense time scales of impacts are being ignored, and the question of monitoring, remediation, and regulation thousands of years into the deep future constitutes a kind of 'uncomfortable knowledge'. Ultimately, I contend that the legacy of the rehabilitated Ranger uranium mine will pose long-term threats to the environment and Mirarr Traditional Owners of the area, and that this slow violence constitutes a kind of unacknowledged environmental disaster, but one which is being disregarded through the active production of ignorance.

Introduction

Slow and seemingly invisible environmental disasters are all around us, but they are particularly acute on Indigenous peoples' territories throughout the world. Indigenous peoples are all too often at the raw end of commodity chains; they are the 'downstream communities' (Plumwood 2008) and generally excluded from the environmental management of mine sites during their operation (O'Faircheallaigh and Corbett 2005). But when lax regulation and intractable rehabilitation challenges leads to mine sites being abandoned (O'Faircheallaigh and Lawrence 2019) it is those Indigenous communities who continue to be haunted (Keeling and Sandlos 2017), as they struggle to keep mining companies and governments accountable for the perpetual care and management of toxic sites on their homelands. In the case of uranium mines, the material reality and longevity of contaminated sites means these injustices are etched into the landscape long into the deep future. These places are emblematic of the 'slow violence' (Nixon 2011) of environmental degradation and the environmental injustices and harms disproportionately affecting Indigenous peoples across the world (Schlosberg 2013).

I explore ideas and practices around 'uncomfortable knowledge' (Rayner 2012) and 'ignorance' (McGoey 2012) and specifically how the extraction of uranium on Indigenous peoples' territories is enabled by a particular kind of ignorance around its immediate and ongoing impacts, as well as impacts long into the deep future. Ignorance, while of considerable focus within science and technology studies and sociology (Gross and McGoey 2015), is not generally well attended to in the environmental politics literature: the role of ignorance in environmental and social injustices is often treated statically, and an information deficit approach tends to dominate (Norgaard 2011, p. 3). But this veil of ignorance is not just about avoiding relevant knowledges of harm; it is also fundamentally concerned with actively producing ignorance, and managing 'uncomfortable knowledge', around places of degradement and potential abandonment, such as uranium mines. The point is that we need to see ignorance as 'regular' or commonplace (Gross and McGoey 2015), and a part of everyday environmental decisions around people and place.

This research makes an important contribution to the literature by attending to this phenomenon through a case study of the planned attempts to clean up, rehabilitate and close the contentious Ranger uranium mine on Mirarr Aboriginal land in the Kakadu region of the Northern Territory in Australia. The Ranger mine officially closed in January 2021 and is currently required to complete rehabilitation works by 2026. While I do not explicitly focus on the impacts of the mine or closure on Aboriginal people, the fact remains that the mine is located on traditional Aboriginal land and any environmental impacts will be most acutely experienced by local Aboriginal communities.

I demonstrate that the idea that a uranium mine will leave no impact, or that its impacts will remain 'contained' – a discourse employed by mining industry and regulatory and monitoring authorities alike – is a fallacy. This requires the astute management of particular kinds of 'uncomfortable knowledge' around environmental impacts, such as seepage from mine tailings and other contaminants, and the careful manufacturing of ignorance regarding the need for the site's perpetual care and maintenance. As scientific contestations discussed here demonstrate, in a monsoonal tropical environment, poised to be impacted by climate change and salt-water inundation (Bayliss *et al.* 2018) where toxic contaminants from tailings and waste-rock are predicted to flow and mix through waterways and groundwater taking contaminants off site, the idea of isolation or containment is a myth. Yet this uncomfortable knowledge is one that has to be managed not only by mining companies, regulators and monitoring agencies, but by environmental NGOs, who are also aware of this conundrum.

I turn now to a discussion of my material and methods. I then outline some of the debates within the literature around uncomfortable knowledge and ignorance. I follow by outlining the empirical case in more detail, with particular attention to forms of ignorance in the uranium mining industry more generally, and specifically at Ranger. This is then explored further through analysing two key challenges regarding the rehabilitation and closure of the Ranger uranium mine: 1) tailings containment and 2) financial security for the rehabilitation and perpetual care and maintenance of the site. My aim is to demonstrate, through this case, that a rich empirical field is to be found in the omissions, gaps and silences of a mine closure process. I illustrate that a multitude of actors are involved in these practices, not just mining companies, but monitoring authorities, NGOs, and even researchers themselves.

Material and methods

The research detailed here was in part produced through a formal research collaboration between myself and the Australian Conservation Foundation (ACF), Australia's largest environmental NGO. The ACF have long played the role of corporate watch-dog at the Ranger site, have campaigned in the Kakadu region for decades, and have a strong presence in the anti-nuclear campaign movement in Australia more generally. The ACF have a formal agreement (the Kakadu Charter) to collaborate on Ranger issues with Gundjeihmi Aboriginal Corporation (GAC). GAC represents the interests of the Mirarr Traditional Owners, including to the wider regional Aboriginal representative body the Northern Land Council (NLC) and the Northern Territory and Commonwealth Governments.

The research collaboration with the ACF resulted in two advocacy reports highlighting the key challenges associated with the rehabilitation and mine closure at Ranger uranium mine, which included ecological rehabilitation, tailings management, climate change, social impacts, stakeholder management, opaque regulatory and assessment regimes, financial security and perpetual care and management of the site.[1] The release of the first advocacy report in May 2019 was timed to coincide with Rio Tinto's Australian annual general meeting, as the main shareholder in ERA, the operator of the mine. While Rio Tinto has long prided itself on being an industry leader on Indigenous engagement, the global controversy that commenced in May 2020 around Rio's destruction of an Aboriginal cultural site of significance at Juukan Gorge for the expansion of its Pilbara iron-ore operations points to the frailty of Rio's corporate reputation. But it also highlights the role of ignorance, or 'not-knowing' in extractive activities, as Rio initially claimed ignorance of the site's significance to Traditional Owners as defence of its actions (Young and Toscano 2020), something not unusual in corporate discourse.

But if not knowing is a common corporate strategy, the act of revealing or uncovering is a classic NGO one (Lawrence 2008): fact-finding missions are the raison d'être of the NGO world. And so, the main activity undertaken as part of this research was built around uncovering and highlighting omissions in ERAs 2018, 2019 and 2020 Ranger Mine Close Plans (RMCP) through detailed reviews. In addition to critical readings of the RMCPs, I also reviewed the 2018 and 2019 assessment reports of the RMCPs by the Supervising Scientist Branch (SSB)- the Commonwealth of Australia's monitoring agency with the sole purpose to monitor the impacts of the Ranger Uranium mine on Kakadu National Park. I assisted the facilitation of a collaboration between ACF and the Environmental Defenders Office (EDO),[2] an Australian-based environmental public law NGO. EDO, in turn, commissioned expert peer-reviews of the 2018 RMCP. On the basis of these formal peer reviews, and discussions with other NGO representatives at the Environment Centre NT and Mineral Policy Institute (MPI) – both small-scale Australia environmental NGOs but with deep knowledge of the Ranger mine – we compiled the first advocacy report, which was co-published by Sydney Environment Institute (SEI) and ACF in 2019 (Lawrence and Sweeney 2019). In 2020 we published a second advocacy report, co-authored by MPI, SEI, ACF and Environment Centre NT (Pepper *et al.* 2020).

As a part of this research, I also spent considerable time updating incident lists (originally compiled by GAC, ACF and the Northern Land Council, which documented environmental incidents and breaches at the Ranger site for the period 1979–2014), with reference to environmental reports from ERA to The Alligator Rivers Region Advisory Committee (ARRAC) for the period 2014–2019. This data collation resulted in a complete list of incidents from 1979 to 2019 and was used in the advocacy report to evidence a total of nearly 1,000 environmental breaches at Ranger. Two field trips were taken, the first was a research scoping trip to Jabiru – the mining town associated with the Ranger mine – which was facilitated by ACF[3] in October 2017, where I informally met with the GAC CEO and other staff. The second was a field trip to Darwin, in October of 2018, where I attended a Science Symposium, held by the Supervising Scientist – the Commonwealth monitoring authority for the Ranger mine – and also met with the Environment Centre NT and the Northern Territory office of the EDO.

While ACF and other NGO staff provided detailed comments on various written drafts of this research, the SSB declined to provide comment, stating 'It is important for SSB to remain outside of the political debate so we won't provide any comments on your paper' (Personal email correspondence, SSB, 11/11/2020). The SSB may have had particular issue with the framing of the research in terms of ignorance (see (Stewart 2015, p. 375), or they may have simply been re-rehashing the common concern by other government agencies and corporations in regard to research on environmental and social justice (both my own and others) (Lawrence and Raitio 2016), namely, that engaged research is neither objective nor neutral and therefore cannot be trusted (see also (Tatz 1984, p. 11). But without a basic fact-check from the SSB on key technical issues, it certainly made the research process more difficult. I have done my best to rely on the expertise of expert peer-reviews, and academic and NGO colleagues with expertise in relevant disciplines (i.e. tailings containment and fresh-water ecology), in addition to a wide reading of the scientific literature. Yet, I must concede, there may well be factual errors contained herein.

One of the key methodological challenges in critically reviewing any assessment document or process is in reading for absence and omissions. But this work, as an activist academic(Lawrence and Raitio 2016), was not just about indiscriminately revealing everything. Because, in some cases, there was an imperative, by NGO representatives and myself, to maintain a level of strategic ignorance around particular kinds of uncomfortable knowledge deemed too difficult to manage. In short, what emerged during this research was my own realisation that the rehabilitation of the Ranger mine was a truly 'wicked problem' (Rayner 2012). As I demonstrate, it is an intractable, highly complex challenge and never likely to fully succeed. Toxic mine tailings cannot possibly be fully contained in a monsoonal tropical environment, and the costs of the rehabilitation and perpetual care and maintenance of the site are so great, that they are generally being ignored. How then, does one frame an advocacy report that seeks to make governments and corporations accountable, when the rehabilitation task before them is politically complex, likely requires a never-ending supply of resources, and is actually, in all likelihood, materially impossible to fully realise? How does one avoid the risk that key actors will simply resign themselves to the hopelessness of the task? One way is to run with the narrative that one has to try and get the best scientific outcome possible, which was the broad narrative underpinning the advocacy reports. But this research now provides an opportunity for a more scholarly reflection of the conundrum at hand: what is materially achievable at the Ranger site is highly contested, and one where ignorance and uncomfortable knowledge play a key role.

Ignorance, uncomfortable knowledge, and its organisation

As Kourany and Carrier argue, thinkers in both epistemology and the philosophy of science – such as Peter Unger and Karl Popper – have been primarily concerned with the limits of knowledge and taken ignorance for granted. Studies of ignorance – or agnotology – flip this approach and seek to explore, analyse and probe ignorance itself (Kourany and Carrier 2020, p. 6). Scholars of ignorance studies have thus turned their attention to the ways in which ignorance is not just the absence of knowledge, but is a social and political achievement in and of itself (Rayner 2012, McGoey 2019) and may be mobilised in order to uphold particular social and environmental injustices. Norgaard has made a similar point: we need to move away from an information deficit approach if we want to really understand why denial, and ignorance, continue to flourish (Norgaard 2011, p. 3).

Crucially, the production of ignorance is also often assembled in close-knit with science (Kirsch 2014). The mining industry often employs various strategies of ignorance, by controlling information, manipulating science in order to limit critique, and 'the strategic promotion of uncertainty and controversy' (Kirsch 2014, p. 130). Both the academic and NGO literature tell countless tales of sites of legacy pollution (Roche and Judd 2016, Sandlos and Keeling 2016), which require ongoing and 'chronic remediation' of contaminants that are not immediately visible. Remediation processes often drag on for decades, consultation processes become overly complex, and no-one knows who is really responsible (Shriver *et al.* 2019).

At work here is often a kind of deliberate or strategic ignorance by corporations (McGoey 2019). But mining legacies are not just stories about strategic and wilful corporate ignorance. If strategic ignorance is 'any actions which mobilise, manufacture or exploit unknowns in a wider environment to avoid liability for earlier actions.[or] future [actions]' (McGoey 2019, p. 3), then there are clear entanglements with 'uncomfortable knowledge', as 'disruptive knowledge. It may be information or understanding that is available to certain parties, but cannot be acknowledged by others.' (Rayner 2012, p. 113). The relationship between ignorance and 'uncomfortable knowledge' and the various actors – including monitoring authorities, NGOs and communities – who navigate this space is slippery and shifting. Strategies concerning uncomfortable knowledge and ignorance are organised, enacted, produced and reproduced through particular social, political and bureaucratic practices (Norgaard 2011) but they are assembled in messy ways. Indeed, even the most technical issues are never devoid of politics (Barry, 2001). To be sure, mining uranium on Aboriginal people's homelands without their consent is certainly a political act, and any related activity, such as scientific research and monitoring of those impacts – however 'technical' – is equally political.

Ignorance at work: uranium mining and the rehabilitation at Ranger

The Ranger mine is surrounded by, but officially excluded from, Kakadu National Park, Australia's largest national park and inscribed on UNESCO's World Heritage register for both its natural and cultural values and properties. This is a monsoonal tropical landscape with rich biodiversity. The Aboriginal Land Rights Act (1976) explicitly excluded Ranger from its veto provisions and the Ranger uranium mine began operations on Mirarr land in 1979. This was a profound social injustice for the Mirarr people, who had consistently opposed uranium mining on their traditional lands (O'Brien 2003). These things combined have made Ranger one of the nation's most scrutinised and contested resource operations, and it has been surrounded by controversy (O'Brien 2003).

Ranger is Australia's longest running uranium mine, and has produced around 11% of the world's uranium for fuelling nuclear power plants (Energy Resources of Australia 2018, p. 1–3). The first ore was milled in August 1981 and the operation was formally opened in October 1981. In 2000, Rio Tinto became the largest shareholder in the Ranger uranium mine. In 2013 the Ranger Agreement was renegotiated between ERA and GAC, which in addition to receiving increased royalty benefits from the mine, also endowed the Mirarr with increased opportunity to participate in the technical and environmental aspects of the mine (Graetz 2015, p. 45). The GAC has, however, continued to be publicly critical of the Ranger operations, most notably in 2013 when a leach tank failed and spilled 1,400,000 L of process slurry on site, something the Mirarr Traditional Owners described as the 'biggest nuclear disaster in Australia ever' (Gundjeihmi Aboriginal Corporation 2013).

Despite this, and the nearly 1000 environmental incidents at the Ranger site collated as a part of this research, the SSB has generally maintained that its monitoring has proved that the mine has had no adverse impacts on the environment: ' … the Supervising Scientist's environmental monitoring has shown that Kakadu National Park has been protected from the effects of mining' (Supervising Scientist 2018b). How is that regulatory and monitoring authorities in Australia are so concerned to uphold the idea that the Ranger mine has no impact on the surrounding environment? Colin Tatz, an anthropologist, asked the same question in 2006: 'On one hand, we have a scientific regulatory agency [SSB] which argues that the water management system at Ranger is the best in the world, and that in no instance has there been an adverse impact on the environment or the human population as a result of "occurrences". On the other, there is a constant record of leaks and spillages ... Which is correct? ' (Tatz *et al.* 2006, p. 14).

History partly explains this phenomenon. When the Ranger mine was being assessed and permitted in the 1970s, it was against the backdrop of 'a highly charged political context' (NGO informant A; see also (O'Brien 2003). First, the mine was permitted and established against the will of the Mirarr people. Second, the mine was established in the face of vehement opposition by the anti-nuclear movement in Australia and a growing environmental movement. Third, the mine site was proposed within the geographic borders of what was to become Kakadu National Park but officially excluded from it (Fox *et al.* 1977). These three issues set the scene for a political and regulatory necessity to commit to the idea of 'containment' (von Sturmer 1984), most likely in order to garner public acceptance of the mine. And so, the Environmental Requirements for the Ranger mine are the strictest of any uranium mine in Australia. Among other things, the mining company is to ensure that:

- " (i) the tailings are physically isolated from the environment for at least **10,000 years;**
- (ii) any contaminants arising from the tailings will not result in any detrimental environmental impacts for at least 10,000 years;" (Clause 11.3, Environmental Requirements, Section 41 Authority)

The Environmental Requirements for the Ranger uranium mine set out the Commonwealth's environmental protection conditions with which ERA must comply: they are embedded in a complex regulatory regime with unclear lines of accountability between different actors. ERA have attempted to articulate through their mine closure plans how the rehabilitation and closure of the Ranger mine will navigate the complex regulatory regime that governs the Ranger operations and how they will comply with the environmental requirements. Yet what neither of these plans reveal is that when the first RMCP was released 'it was initially unclear who held ultimate authority for assessing, approving and enforcing that plan, with the long list of regulatory advisory bodies adding to the convolution'. As Lea et al. argue, it was a case of 'everyone and no one [being] responsible' and was not until the GAC worked to have amendments made to NT legislation, that some clarity was achieved (Lea *et al.* 2018, p. 312).

Weak and complex regulatory regimes arguably serve the interests of capitalism. But the lens of regulatory capture alone does not do this case justice. The role of ignorance also has a key role in upholding capitalism and settler colonialism. This is illustrated through the discursive work that containment performs: after all, we can only continue to legitimately approve uranium mines (or, any mines) on Indigenous lands so long as we uphold the idea of containment and ignore the messy actualities of toxic legacies long into the deep future.

Indeed, the idea that a uranium mine can operate without impacting on the environment is crucial to the broader narrative that nuclear power is a clean source of energy: an alternative to fossil fuels in our war against climate change. This is a popular narrative in Europe, with mainstream politicians in countries such as Sweden, arguing for more nuclear reactors as a way to reduce climate change, claiming Sweden's energy 'based on nuclear, hydro-power and wind-power, is the cleanest in the whole of the EU' (Kristersson and Sabuni 2020). This discourse is also apparent at the local scale, on the very sign outside the Ranger mine site itself, which states that 'nuclear energy . . . can help alleviate climate change as it produces much lower emissions than other electricity sources'.[4]

Yet there is very little public debate around the impacts of mining uranium as a raw material, and a general lack of attention to the links between the impacts of uranium mining to the 'clean energy' produced by nuclear power plants. Indeed, young people between 18 and 34 in Australia are the least informed about nuclear power, and 'some [aren't] even sure there is a connection between nuclear power and uranium' (Patrick 2020, p. 118). As Burke contends '[d]ebates about nuclear energy focus around electricity prices, renewable energy and climate change, but rarely touch on the ecological and social impact of uranium mining' (Burke 2017, p. 7). These debates thus reveal a common way in which ignorance is maintained: diversions or distractions (Rayner 2012, p. 118) are created – that is, by focussing on climate change debates – and the impacts of uranium mining remain hidden from view.

Sometimes the real utility of ignorance is in 'knowing what not to know' (McGoey 2012, p. 554). At the time of writing, the Minerals Council of Australia (MCA), representing the Australian mining industry, is lobbying to have crucial Commonwealth protections and scrutiny around uranium mining removed from the Environment Protection and Biodiversity Conservation Act 1999, Australia's key federal environment legislation. In their formal submission, the MCA make no mention of the intractable challenges involved in rehabilitating uranium mines[5] yet Australia has a notorious record when it comes to mine rehabilitation more generally. The mining industry's optimism in its ability to mine fragile ecosystems, and then later 'restore' these to their original state (Beckett and Keeling 2019), is rarely backed up with material examples. At the 2018 Senate Inquiry into mining rehabilitation, the Minerals Council of Australia could not highlight a single case of a successfully rehabilitated large-scale mine site (Young 2018). This failure is underpinned by a systematic lack of transparency, accountability and financial security around rehabilitation and post-closure frameworks, which has particularly acute impacts for already vulnerable Indigenous communities (O'Faircheallaigh and Lawrence 2019). In fact, abandonment, asset transfers, and indefinite care and maintenance arrangements are more common corporate exit techniques than orderly mine closure. Consequently, there are over 50,000 abandoned mines in Australia and while many of these mines may be small single shaft mines, others are not. In many cases, no rehabilitation efforts have even been made (Roche and Judd 2016) and even when they are, they may re-emerge decades later with new environmental impacts, such as those demonstrated by the rehabilitation failure of the Rum Jungle uranium mine in the Northern Territory of Australia (Mudd 2010).

I now turn to explore the production of ignorance through two key issues at the Ranger mine site: 1) tailings containment and 2) financial security and perpetual care and maintenance of the rehabilitated site. The former speaks directly to one of the main environmental requirements and is dealt with in ERA's mine closure plans and SSB's Assessment reports, the latter is almost entirely absent, but is crucial to realising the immediate rehabilitation and its long-term success.

Tailings containment

Mine tailings at Ranger consist of toxic residues produced from processing uranium from the mined ore, which contain radionuclides, heavy metals and chemicals. Despite the requirement to isolate the tailings from the surrounding environment for at least 10,000 years, the mine has struggled to fully comply with this requirement, even during operations. In 2009, the Supervising Scientist confirmed that the tailings dam was leaking 100,000 cubic litres of contaminants a day into rock fissures beneath Kakadu (Murdoch 2009). This stands in stark contrast to the original predictions by the mining company in the late 1970s, which would mean that tailings seepage should be around 1780 cubic metres by year 2010, a gross underestimation given the actual seepage in 2009 at 56 times that rate.[6] The resulting plume of contamination beneath the tailings dam remains a significant remediation challenge for ERA and has been noted by the Supervising Scientist as a rehabilitation risk (Supervising Scientist 2018a). In the 2020 RMCP, it is not clear if ERA intend to remediate the contamination plume at all (Energy Resources of Australia 2020, Pepper *et al.* 2020). For environmental NGOs concerned with Ranger, this, and other breaches at the mine, illustrate 'the far-reaching self-deception that contamination can be contained within the dotted lines' (NGO Informant B).

During the rehabilitation, mine tailings from the tailings dam will be placed in the mined-out pits. However, at the time of writing, there were ongoing NGO concerns that the floor and walls of the tailings dam, which contain tailings residue, will not be removed and buried in the pits by ERA, but will be left in situ, and that this would increase the risk of contaminant leaching (Pepper *et al.* 2020).The pits are planned to be mostly backfilled with radioactive waste rock and capped with a final surface cover of clean or non-mineralised rock. In a tropical monsoonal environment, such as in Kakadu, even after the rehabilitation has been completed, the mine site will continue to pollute heavy metals and chemicals into the surrounding environment. The main pathways for this pollution are via contaminant leaching from the surface waste rock used to cover the tailings and from the tailings and brine buried in the mined-out pits. When water moves through the waste rock, tailings and brine it will mobilise contaminants, such as heavy metals and chemical residues. In rehabilitation terms, uranium, magnesium and sulphate are the solutes of most concern.[7]

The SSB has been researching and establishing 'guideline values', which set theoretical thresholds for levels of individual contaminants that aquatic species can tolerate whilst ensuring that 99% of the ecology is protected. These will inform rehabilitation standards for the Ranger mine. This exercise essentially asks: what are the acceptable contaminant thresholds/levels for sensitive species? In short, this is not a question of complete containment or isolation, but of establishing guidelines for tolerable rates of contaminant leaching. Despite the growing knowledge in the field of ecotoxicology, many challenges and uncertainties remain (Breitholtz *et al.* 2006). These are compounded by the rehabilitation of the Ranger mine, and the climate change impacts of salt-water inundation on aquatic species in Kakadu (Humphrey *et al.* 2018).

There have been contestations over what these guideline values should be, in relation, for example, to dissolved concentrations of uranium and manganese in surface water after rehabilitation works are completed. For uranium, the SSB Rehabilitation Standard proposes 2.8 ug/L, yet ERA has proposed a much higher acceptable limit of U 17 ug/L. Similarly, for manganese, the SSB Rehabilitation Standard proposes 75 ug/L, yet ERA has proposed a much higher acceptable limit of Mn 500 ug/L.[8] It is unclear if these disputes have been resolved. The final negotiation of rehabilitation standards is yet to be agreed upon, and it is unclear who will yet have the final say: the mining company or the SSB.

But this is only one part of the equation. The problem is that knowing what theoretical levels of contaminants the current surrounding environment can *tolerate* is a very different question than asking what contaminant levels will *actually* be coming off the mine site. But while the former can known (or, at least, estimated) by performing real-time experiments in the laboratory, the latter can only be modelled. This is a space constituted by multiple uncertainties and unknowns. First, and as noted above, there is a significant complexity and considerable unknowns in terms of the contaminant levels coming off site. Second, there is further uncertainty as to how contaminants from various pathways will behave when they mix. No complete modelling has yet been released by ERA of contaminants coming off site, nor how they will interact with the surrounding environment as various contaminant pathways from waste-rock and mine tailings mix, potentially increasing in toxicity as they do so (Supervising Scientist 2018a, p. 48). SSB have identified contaminant transport modelling as the key area of concern in their assessment of the 2019 RMCP and raised a red flag that modelling may not be completed by ERA in time for rehabilitation works to proceed (Supervising Scientist 2019, p. iv). ERA have stated in both the 2019 and 2020 RMCPs that this modelling is currently being undertaken (Energy Resources of Australia 2019, p. 11–15) (Energy Resources of Australia 2020, p. 5–233) but it is unclear when it will be finalised, or if it will be publicly released. In the words of the SSB, this is a 'Key Knowledge Need', or, as Donald Rumsfeld would say, 'a known unknown'. Third, any modelling also runs into the problem of peak contaminant loads. Peak contaminant loads concern the time it takes for contaminants on site (at the time of mine closure, in the form of tailings and waste rock) to reach their peak off site. Contaminants from waste rock will not reach peak loads downstream until around 300 years post-closure and tailings will not reach their peak loads downstream until 1000s of years post-closure. In other words, there are enormous time-lags to consider, and if modelling is incorrect, we may not know it for hundreds or thousands of years.

This raises a critical remediation challenge: what would ERA actually be able to do if the flow of contaminant mixtures off-site exceeded safe thresholds? Or, if variables outside the model (unknown unknowns) increase the toxicity of contaminant mixtures coming off-site and the surrounding environment is adversely affected? These are important questions, but ones which are not being explicitly asked by the SSB or ERA. Instead, a narrow focus on guideline values and the modelling is 'displacing' more long-term concerns around accountability and corporate capacity. The accurate modelling of contaminants (if achieved) is important, but it is also an act of distraction. It 'substitutes a more manageable surrogate' (Rayner 2012, p. 120) – that is, the establishment of guidelines and models – so as to distract from the uncomfortable knowledge around responsibility long into the deep future, and thus avoid the more fundamental questions: what corporate entity is likely to be held accountable in 50, 100, 1000 or 10,000 years if the models are wrong?

Indeed, this uncomfortable knowledge concerning the (in)capacity of mining companies to respond to remediation issues post-closure emerged during the research and writing of the first advocacy report. As one NGO representative expressed:

> I'm worried that the environmental impact of the rehabilitated site will actually be worse than it currently is. Now, it is an operational site and ERA have a capacity to respond to issues. But once Ranger is rehabilitated and closed they [ERA/Rio Tinto] won't have that same capacity, but I'm worried if we say that it will be too negative . . . (NGO Informant B)

In regard to the question of tailings containment, another NGO representative reflected:

> It's a fiction to talk about the containment or isolation of the tailings. Water will be flowing through the tailings and taking with it contaminants off-site, yes. So it's really about the rate of water flow and levels of contamination. But while absolute containment is a fiction the need for best environmental practise and protection is a fact. We have fought hard to uphold this standard and ambition because we want the best possible outcome for the environment. (NGO Informant A).

This reveals the complex navigations that all parties to environmental conflicts engage in. Mining companies, government monitoring authorities and environmental organisations alike are required to manage and position themselves in regard to particular kinds of uncomfortable knowledge. But it also illustrates the kinds of narratives and strategies civil society organisations may need to adopt to get traction within particular policy spaces, even if that means upholding an unrealisable ambition around rehabilitation.

Financial security, and perpetual funds and management

Given the environmental requirement that tailings be isolated for 10,000 years, financial security of the rehabilitation itself, as well as a perpetual fund and management plan, are both key to ensuring its success. Estimated rehabilitation costs at Ranger have increased multiple times. In April 2016 Rio Tinto made a commitment to provide a credit facility of 100 USD million to ERA, boosting the rehabilitation budget to over 500 USD million. But this was insufficient and in December 2018, the predicted clean-up bill was increased from the 512 USD million estimate to 808 USD million – 296 USD million more than ERA's initial estimate.[9] In February 2019, this amount was further increased to 896 USD million.[10]

Yet, none of these figures have ever included costs for the perpetual care and maintenance of the site – they have only ever included the costs of rehabilitating the site up until closure. Mine rehabilitations are notoriously tricky, often fail and commonly require remediation works at a later point (Lamb *et al.* 2015). This issue has long been of concern to many, including the Mirarr Traditional Owners. In 2005, the GAC recommended that ERA be 'required to establish a fund in perpetuity that can be used to maintain and monitor the rehabilitated area and if necessary repair any of the rehabilitation works that fail'.[11] Similarly, in 2008, ACF formally raised concerns over the lack of monitoring funds, and the risk that cost-shifting could see this become a Commonwealth liability and ultimately a tax-payer burden (Australian Conservation Foundation 2008). There is currently no perpetual management plan in place, which would include monitoring and a perpetual fund for remediation works. The 2019 and 2020 RMCPs state that monitoring will take place for 25 years (Energy Resources of Australia 2019, p. 12–13) (Energy Resources of Australia 2020, p. 10–13) but this figure appears to born out of corporate convenience, rather than in relation to any actual site needs.

In 2008, ACF and EDO modelled a cost estimate for monitoring groundwater, radiation and surface water at the Ranger site for 1,000 years post-closure at a sum of between 19 billion and 387 trillion AUD. The scale of these figures is difficult to comprehend. But they clearly outweigh any value or benefits produced by the mine during its operational life, raising the issue as to whether mining is actually a profitably enterprise, or, how these costs and benefits are inequitably distributed across the life of the mine, across different populations and across different geographies: foreign shareholders and domestic governments receive the overwhelming majority of benefits (via dividends and royalties) while benefits to local Indigenous communities are limited at best. Furthermore, the total costs of mining are born not by the mining company, but by the environment, tax-payers and, ultimately, Indigenous communities (O'Faircheallaigh and Lawrence 2019). Crucially, while the benefits are extracted during the operational life of the mine, many of the costs and impacts are often borne far into the future. Modelling the future costs of monitoring is one way to make this injustice visible. But demanding that these costs be borne by companies about to embark upon a complex rehabilitation project – with precarious funding and critical minority shareholders (such as is the case of Rio Tinto at the Ranger site) – may create its own kinds of risks. During the research and writing of the first Ranger report published by ACF and SEI, this conundrum became apparent, as one NGO representative reflected:

> This rehabilitation will be complex and costly and we need to think about if we want to fully detail these monitoring costs right now. Will we be taken seriously or do we risk scaring them [ERA/Rio Tinto] off, and them walking away from the rehabilitation all together? (NGO Informant A)

The modelled monitoring costs were excluded in the first report, although the need for increased financial certainty was an important theme throughout. This illustrates how knowledge around the gravity of the financial costs of monitoring may be too uncomfortable to acknowledge publicly, not only for mining companies, but also for civil society actors and academics seeking to hold corporations and governments accountable and committed to the rehabilitation task before them.

Conclusion

The rehabilitation issues facing the Ranger mine are numerous. I have provided a snapshot of just two of them. In the case of the first, the myth of complete containment of tailings requires the management of uncomfortable knowledge that contaminants will in fact leach into the surrounding environment, as well as significant uncertainty over what those contaminant levels will be, and how they will behave once they mix with other contaminant sources, including waste rock. No-one has the benefit of hindsight to know how a large-scale rehabilitated uranium mine site behaves long into the deep future. Yet regulators and the mining industry avoid this, by attending instead to modelling of contaminants and scientific debates over environmental thresholds for contaminants. Notably, the discourse of containment is not unique to the management of uranium mine tailings: it is also mobilised at the back-end of the nuclear cycle in regard to nuclear waste and does a specific kind of 'rhetorical boundary work' as it seeks to set both discursive and material boundaries around 'leakages' (Kinsella 2001). The narrative of containment is thus one that travels along the nuclear cycle as it seeks to uphold the idea that toxic materials can, in fact, be isolated.

But this narrative of containment also requires ignoring any need for nuclear semiotics, which appear to be completely absent from any discussions at Ranger. After all, if the impacts are contained, why should both humans and non-humans, long into the deep future, need to know anything about contaminants? Yet the material reality is that sinking bores for extraction from groundwater will be prohibited post-closure,[12] which may well be a prudent public health measure, but it is in direct contradiction to the narrative of the 'containment' of the mine's impacts. Moreover, while the RMCP details how Mirarr concerns around the final landform have been integrated into closure criteria (e.g. regarding ground preparation, plant species etc) it is unclear if, and how, Mirarr aspirations and concerns around groundwater contamination have been managed or responded to. It seems these groundwaters on the 'resource frontier' constitute particular kinds of 'shadow waters': invisible, seemingly disconnected from surface waters, and with little or no regard for Indigenous epistemologies (McLean *et al.* 2018).

In the case of the second issue, the question of how remediation will be undertaken or financed if rehabilitation works fail is completely absent from any of ERA's Mine Closure Plans. There is no mention of a perpetual care and management plan or how such a plan would be funded: this omission is testament to the strategic use of ignorance for maintaining social and environmental injustices. ERA and Rio Tinto can potentially walk away after the end rehabilitation date in 2026, leaving significant uncertainty as to who is responsible for monitoring, for any remediation works arising, and how such activities are to be financed. In such case, these costs would, ultimately, become externalities to be borne by the public and local communities.

I conclude with a more general observation around the reluctance of ERA to clearly articulate and acknowledge their own knowledge gaps – or ignorance – in their Mine Closure Plan. This is a kind of meta-ignorance: a strategic ignorance of their own ignorance. This is achieved by way of distraction: all three iterations of the RMCPs (2018, 2019 and 2020), are enormously complex documents, but all fail to clearly outline the key challenges ahead. This constitutes what Lea et al call 'slow administrative violence' through 'paperfare': a battle waged by corporations and governments by way of relentless bureaucratic requirements made of Aboriginal organisations, like the GAC, to respond to multiple iterations of complex documents, of which mine closure plans are just one of many (Lea *et al.* 2018, p. 316). It is also an example *par excellence* of what Heimer calls 'sequestered knowledge' (Heimer 2012): piles of paper and reports are produced creating the illusion of knowledge, whilst avoiding key knowledge deficiencies. Indeed, the RMCP's gives the overwhelming impression that everything is understood and on track. This is simply not the case.[13] In their 2018 assessment of the RMCP, the Office of the Supervising Scientist concluded the following: 'The MCP . . . . does not yet provide sufficient evidence to demonstrate that the current plan for rehabilitation of the Ranger mine site will achieve the required ER's [Environmental Requirements]' (Supervising Scientist 2018a, p. viii). Again, in their 2019 assessment, SSB expressed similar concern, particularly in regard to the lack of certainty around contaminants coming off site and the risks they pose to the downstream environment (Supervising Scientist 2019, p. iv–v).

Veiled as they are in the bland language of bureaucracy, these are alarming assessments by a key Commonwealth agency. They also directly contradict the narrative that toxic materials can be contained. Yet there has been no formal response by the NT or Commonwealth Governments, no review of the regulatory processes, and no obvious increase in pressure on ERA. But it would not, after all, be the first time a Government remained wilfully ignorant of agency reports bearing red flags about corporate complacency. It is not necessarily the absence of knowledge that forebodes environmental and social disasters – however slow they may be – but the wilful production and maintenance of ignorance around them. If we are to appreciate the complexity of how social and environmental injustices are produced, and reproduced, we must attend to ignorance as more than just an absence of knowledge or failure to know. It is actively constituted by a suite of activities that seek to distract and displace uncomfortable knowledges for other, more manageable, ones.

Notes

1. Note the research reported in this published research does not deal with all the key issues detailed in the original advocacy report, but future scholarly publications will, particularly regarding social impacts.
2. The author has provided pro-bono peer reviews to the EDO on the social impacts of resource developments and had an existing relationship with them.
3. ACF provided an introduction to the Gundjeihmi Aboriginal Corporation (GAC).
4. Fieldwork photograph, Lawrence, October 2017.
5. See Minerals Council Of Australia Submission To The Independent Review Of The Environment Protection And Biodiversity Conservation Act 1999.
6. In the original Fox commission into the impacts of Ranger the mining company estimated that seepage at year 10 (year 1990) would be 470 m3 and at year 20 (year 2000) would be 890 m3 (pg 99, Fox report). Based on the original estimation that seepage rates would double every 10 years, the 2010 seepage rate should be approximately 1780 m3.
7. I acknowledge an anonymous academic for contributions to the scientific description of the tailings.
8. For SSB's Rehabilitation Standards see https://www.environment.gov.au/science/supervising-scientist/publications/ss-rehabilitation-standards. For ERA's proposed closure criteria see page 6–19, in ERA (2018) Mine Closure Plan, Issued Date: May 2018. Revision #: 0.18.0.
9. https://www.abc.net.au/news/2018-12-11/ranger-uranium-mine-rehabilitation-cost-blowout-jabiru/10601696
10. https://www.australianmining.com.au/news/era-increases-rehab-costs-for-ranger-uranium-project/
11. Gundjeihmi Aboriginal Corporation, Submission to House of Representatives Standing Committee on Industry and Resources Inquiry into Developing Australia's Non-Fossil Fuel Energy Industry, May 2005, pg 12.
12. While ERA and SSB maintain that the Ranger mine will not leave an impact on the Kakadu region, the 2019 MCP explicitly prohibits the sinking of bores on site for groundwater extraction, other than for monitoring purposes.
13. Planned publication of additional research will demonstrate that other key issues, including social impacts, add further weight to these conclusions.

Acknowledgments

I would like to extend my sincerest thanks to the following organisations and people for their generous support, constructive feedback and discussions throughout the research and writing process: the Australian Conservation Foundation, Gundjeihmi Aboriginal Corporation, Mineral Policy Institute, Environment Centre Northern Territory, Jon Altman, Else Grete Broderstad, Marnie Graham, Richie Howitt, Cathy Howlett, Arn Keeling, Deanna Kemp, Jess McLean, Fiona Miller, Kari Norgaard, Rosemary O'Donnell, Ciaran O'Faircheallaigh, Susan Park, Margaret Raven, Michelle St Anne and the Sites of Violence Project, and the two anonymous referees. I would especially like to thank Dave Sweeney for including me in his advocacy work at Ranger and for his collegiality and generosity throughout the research process. While the OSS declined to comment on written drafts of this article, I thank them for permitting me to attend their Science Symposium in 2018. The views contained herein remain the sole responsibility of the author.

Disclosure statement

No potential conflict of interest was reported by the author(s).

Funding

This research was approved by the Macquarie University Human Research Ethics Committee (HREC (Human Sciences & Humanities)) Reference No: 5201701099 and funded by the Norwegian Research Council under the auspices of the IndKnow project [grant number 288598], The Swedish Research Council for Sustainable Development [grant number 1496401] and the REXSAC project (NordForsk).

References

Australian Conservation Foundation, 2008. *Ranger uranium mine: liability of post-closure tailings management.* Australian Conservation Foundation.

Barry, A., 2001. *Political machines: governing a technological society.* Barry- The Athlone Press, London

Bayliss, P., *et al.*, 2018. Assessing sea level-rise risks to coastal floodplains in the Kakadu Region, northern Australia, using a tidally driven hydrodynamic model. *Marine & Freshwater Research*, 69 (7), 1064–1078. doi:10.1071/MF16049

Beckett, C. and Keeling, A., 2019. Rethinking remediation: mine reclamation, environmental justice, and relations of care. *Local Environment*, 24 (3), 216–230. doi:10.1080/13549839.2018.1557127

Breitholtz, M., *et al.*, 2006. Ten challenges for improved ecotoxicological testing in environmental risk assessment. *Ecotoxicology and Environmental Safety*, 63 (2), 324–335. doi:10.1016/j.ecoenv.2005.12.009

Burke, A., 2017. *Uranium*. Cambridge:Burke - Polity Press.

Energy Resources of Australia, 2018. *2018 ranger mine closure plan*. Energy Resources of Australia.

Energy Resources of Australia, 2019. *2019 ranger mine closure plan*. Energy Resources of Australia.

Energy Resources of Australia, 2020. *2020 ranger mine closure plan*. Energy Resources of Australia.

Fox, R.W., Kelleher, G., and Kerr, C., 1977. Ranger uranium environmental inquiry: second report. Canberra: AGPS, 415.

Graetz, G., 2015. Ranger uranium mine and the Mirarr (Part 2), 2000–2014:'A risk to them is a risk to us'. *The Extractive Industries and Society*, 2 (1), 142–152. doi:10.1016/j.exis.2014.10.005

Gross, M. and McGoey, L., 2015. *Routledge international handbook of ignorance studies*. Routledge. London and New York: Gross and McGoey.

Gundjeihmi Aboriginal Corporation, 2013. One million litres of uranium waste spills into Kakadu [Press release]. Available from: http://www.mirarr.net/news_items/nit-one-million-litres-of-uranium-waste-spills-into-kakadu

Heimer, C.A., 2012. Inert facts and the illusion of knowledge: strategic uses of ignorance in HIV clinics. *Economy and Society*, 41 (1), 17–41. doi:10.1080/03085147.2011.637332

Humphrey, C.L., Bishop, K.A., and Dostine, P.L., 2018. Vulnerability of fish and macroinvertebrates to key threats in streams of the Kakadu Region, northern Australia: assemblage dynamics, existing assessments and knowledge needs. *Marine & Freshwater Research*, 69 (7), 1092–1109. doi:10.1071/MF16175

Keeling, A. and Sandlos, J., 2017. Ghost towns and zombie mines: the historical dimensions of mine abandonment, reclamation, and redevelopment in the Canadian North. *In*: S. Bocking and B. Martin, eds. *Ice blink: navigating Northern environmental history*. University of Calgary Press, 377–420. Calgary, Alberta, Canada: Sandlos and Keeling.

Kinsella, W.J., 2001. Nuclear boundaries: material and discursive containment at the hanford nuclear reservation. *Science as Culture*, 10 (2), 163–194. doi:10.1080/09505430120052284

Kirsch, S., 2014. *Mining capitalism: the relationship between corporations and their critics*. University of California Press.

Kourany, J. and Carrier, M., 2020. *Science and the production of ignorance: when the quest for knowledge is thwarted*. Cambridge, Massachusetts: MIT Press.

Kristersson, U. and Sabuni, N., 2020, July 26. Sverige behöver en ny kärnkraftslagstiftning. *Dagens nyheter*. Available from: https://www.dn.se/debatt/sverige-behover-en-ny-karnkraftslagstiftning/

Lamb, D., Erskine, P.D., and Fletcher, A., 2015. Widening gap between expectations and practice in Australian minesite rehabilitation. *Ecological Management & Restoration*, 16 (3), 186–195. doi:10.1111/emr.12179

Lawrence, R., 2008. NGO campaigns and banks: constituting risk and uncertainty. *In*: *Hidden hands in the market: ethnographies of fair trade, ethical consumption, and corporate social responsibility*, 241–269. Bingley, United Kingdom: Emerald Group Publishing Limited.

Lawrence, R. and Raitio, K., 2016. Academia and activism in Saami research: negotiating the blurred spaces between. *In*: A.-L. Drugge, ed. *Ethics in indigenous research: past experiences-future challenges*. Umeå: Vaartoe-Centre for Sami Research, 117–136.

Lawrence, R. and Sweeney, D., 2019. *Unfinished business: Kakadu needs a new approach to cleaning up an old mine*. Melbourne: Sydney Environment Institute and Australian Conservation Foundation.

Lea, T., Howey, K., and O'Brien, J., 2018. Waging paperfare: subverting the damage of extractive capitalism in Kakadu. *Oceania*, 88 (3), 305–319. doi:10.1002/ocea.5203

McGoey, L., 2012. The logic of strategic ignorance. *The British Journal of Sociology*, 63 (3), 533–576. doi:10.1111/j.1468-4446.2012.01424.x

McGoey, L., 2019. *The unknowers: how strategic ignorance rules the world (Economic controversies)*. London: Zed Books Ltd.

McLean, J., *et al.*, 2018. Shadow waters: making Australian water cultures visible. *Transactions of the Institute of British Geographers*, 43 (4), 615–629. doi:10.1111/tran.12248

Mudd, G.M., 2010. The environmental sustainability of mining in Australia: key mega-trends and looming constraints. *Resources Policy*, 35 (2), 98–115. doi:10.1016/j.resourpol.2009.12.001

Murdoch, L., 2009, Mar 13. Polluted water leaking into Kakadu from uranium mine. *Sydney Morning Herald*. Available from: https://www.smh.com.au/national/polluted-water-leaking-into-kakadu-from-uranium-mine-20090312-8whw.html

Nixon, R., 2011. *Slow violence and the environmentalism of the poor*. Cambridge, Massachusetts: Harvard University Press.

Norgaard, K.M., 2011. *Living in denial: climate change, emotions, and everyday life*. Cambridge, Massachusetts: MIT Press.

O'Brien, J., 2003. Canberra yellowcake: the politics of uranium and how Aboriginal land rights failed the Mirrar people. *Journal of Northern Territory History*, (14), 79.

O'Faircheallaigh, C. and Corbett, T., 2005. Indigenous participation in environmental management of mining projects: the role of negotiated agreements. *Environmental Politics*, 14 (5), 629–647. doi:10.1080/09644010500257912

O'Faircheallaigh, C. and Lawrence, R., 2019. Mine closure and the Aboriginal estate. *Australian Aboriginal Studies*, (1), 65.

Patrick, A., 2020. Young target of nuclear power push. *The Australian Financial Review*, 10.

Pepper, M., *et al.*, 2020. *Closing ranger, protecting Kakadu*. Melbourne: Australian Conservation Foundation, Sydney Environment Institute, Minerals Policy Institute, and Environment Centre Northern Territory.

Plumwood, V., 2008. Shadow places and the politics of dwelling. *Australian Humanities Review*, 44 (2008), 139–150.

Rayner, S., 2012. Uncomfortable knowledge: the social construction of ignorance in science and environmental policy discourses. *Economy and Society*, 41 (1), 107–125. doi:10.1080/03085147.2011.637335

Roche, C. and Judd, S., 2016. *Ground truths: taking responsibility for Australia's mining legacies* Available from: http://www.mpi.org.au/wp-content/uploads/2016/06/Ground-Truths-2016-web.pdf

Sandlos, J. and Keeling, A., 2016. Toxic legacies, slow violence, and environmental injustice at Giant Mine, Northwest Territories. *Northern Review*, (42), 7–21–27–21.

Schlosberg, D., 2013. Theorising environmental justice: the expanding sphere of a discourse. *Environmental Politics*, 22 (1), 37–55. doi:10.1080/09644016.2013.755387

Shriver, T.E., *et al.*, 2019. Industrial pollution and acquiescence: living with chronic remediation. *Environmental Politics*, 1–20. doi:10.1080/09644016.2019.1654239

Stewart, A., 2015. Managing with ignorance. *Routledge International Handbook of Ignorance Studies*, 370.

Supervising Scientist, 2018a. *Assessment report of 2018 ranger mine closure plan*. Darwin: Commonwealth Government of Australia.

Supervising Scientist, 2018b. *The role of the supervising scientist in rehabilitating the ranger uranium mine*. Darwin: Commonwealth Government of Australia.

Supervising Scientist, 2019. *Assessment report of 2019 ranger mine closure plan*. Darwin: Commonwealth Government of Australia.

Tatz, C., 1984. Monitoring social impact. *In*: Australian Institute of Aboriginal Studies, ed. *Aborigines and uranium: consolidated report to the minister for aboriginal affairs on the social impact of uranium mining on the aborigines of the Northern Territory*, 1–19 Canberra: Australian Government Publishing Service

Tatz, C., *et al.*, 2006. Aborigines and uranium: monitoring the health hazards. *Australia: AIATSIS research discussion paper*.

von Sturmer, J., 1984. A critique of the fox report. *In*: Australian Institute of Aboriginal Studies, ed. *Aborigines and uranium: consolidated report on the social impact of uranium mining on the aborigines of the Northern Territory*. Canberra: Australian Government Publishing Service, 20–103.

Young, E., 2018. Mine rehab in WA is the pits: inquiry finds few success stories. *WA today*. Available from: https://www.watoday.com.au/national/western-australia/mine-rehab-in-wa-is-the-pits-inquiry-finds-few-success-stories-20180308-h0x7w9.html

Young, E. and Toscano, N., 2020, June 6. 'Incomprehensible': how Rio Tinto reduced 46,000 years of history to rubble. *Sydney Morning Herald*.

Disaster? No surprise

Christine J. Winter

ABSTRACT

I argue environmental disasters are an inevitable outcome of the underdevelopment of Western philosophy. Human induced environmental disasters are unfortunate, an unintended consequence of focusing on economic maximization in a culture without comprehensive, clearly accepted, and operationalized frameworks for environmental and intergenerational justice. However, they are of no surprise to Indigenous Peoples relegated to margins of political influence. Grounding this claim, I examine the New Zealand High Court case Greenpeace and Te Whanau-ā-Apanui vs the Minister for Energy. Case documents afford comparisons between the complex philosophic, legal, and cultural framework of Māori, that protect human and nonhuman from human-induced environmental disaster, and the dominant Anglo framings. I observe a philosophical deficit underpinning the legal and political system of Aotearoa that impedes environmental intergenerational justice. These philosophic foundations, intentionally or not, *promote* environmental disaster and – nationally and internationally – claim a dominant position over Indigenous systems.

Introduction

Despite 'green tape' – environmental protection legislation and regulation – disasters recur. When the Exxon Valdez tanker spilled oil into the pristine waters of the Arctic in 1989 the world was aghast. The 2010 Deepwater Horizon explosion, subsequent fire and widespread oil spills in the Gulf of Mexico again held audiences in numbed incomprehension. There are of course many other examples, from the Nigerian coastal margins, to Bhopal and the Amazon Basin, Chernobyl and Fukushima and the intensification of fire seasons, droughts, typhoons/hurricanes and floods. While each disaster is unique, I am arguing there is a common core – a philosophy that *supports* environmental degradation. I propose environmental protection is hampered by the liberal political philosophy that undergirds the politics and institutions of western democracies and international organizations (Barnett and Finnemore 2005, Tallberg *et al.* 2020). Moreover, it also frames the structures and instruments of environmental and disaster legislation.

Specifically, I argue human induced environmental disasters (disasters) are an *inevitable outcome* of western political philosophy and that laws, regulations and conventions based in that philosophy are unable to adequately forestall nor respond to them. It is a systemic dynamic that is incompatible with eco-systems based precaution and with thinking compatible with intergenerational environmental justice (Winter 2020). While liberal philosophy may seem to be a universal frame, and therefore one within which the world must work, I examine the ways in which this philosophy may be considered neither universal nor comprehensively just.

To demonstrate the philosophic cleavages I will focus on Aotearoa New Zealand (Aotearoa).[1] I use an example with the *potential* for disaster – deep-sea oil prospecting and drilling off the East Coast of the country. A contrast will be drawn between the Māori practice and protocols of intergenerational environmental justice (kaitiakitanga)[2] crafted to forestall environmental disaster and the government's approach which compartmentalizes risk. In making the intercultural, inter-philosophic comparisons I draw from Te Whānau-ā-Apanui's (Apanui) court testimony and that of the Crown and High Court of New Zealand during the case heard in 2011–2012: CIV-2011-485-1897 [2012] NZHC 1422, Greenpeace of New Zealand Incorporated and Te Runanga o Te Whanau-a-Apanui vs. The Minister for Energy and Resource and Petrobas International Braspetro BV (the Case).

I start with a definition of disaster and approaches to disaster mitigation to establish the base-line epistemological position for international disaster regulation and its liberal philosophic underpinnings. After a brief introduction to philosophy, politics, and colonialism as they have shaped the settler state of Aotearoa I turn to the Case. In addition to the Judgement itself, I examined the extensive materials in the affidavits of the Minister of Energy (the Minister) and his bureaucrats including tender documents and processes, email and meeting records, and bureaucrats' correspondence. These form the corpus for a detailed content analysis to evaluate the New Zealand government's philosophic ground and that of the law. They expose the protocols, processes and the legal framework under which the Minister operates. The contrast is then drawn from Apanui's single affidavit, that of Te Kou Rikirangi Gage. It provides a grounding in Māori philosophic approaches to environmental protection. This affidavit clearly outlines Apanui tikanga – the ethics, philosophy, knowledge and the principles for deciding the *right thing to do*. It explains protocols and processes that protect the water, plant and animal life of the sea, sea floor and coastline, uphold Apanui spiritual values, and support the peoples of the coast. It clearly outlines the procedures Apanui define as appropriate in determining whether or not to open the Raukumara basin to mineral exploration and drilling. The discussion that follows compares the epistemological and ontological foundations of kaitiakitanga and that of the dominant governing systems of Aotearoa. It demonstrates that tikanga is focused on protecting the environment from disaster, while the state governance systems make disaster unsurprising.

What is a 'disaster'?

While we may have an intuitive response to the question 'What is a disaster?', defining disasters seems less straightforward. I suggest it is not straightforward for two reasons: definitions vary by author and over time; and because the kinds of human-induced disasters are variable. Generally, however, definitions propose a disaster has temporal and spatial components and impact human wellbeing. Temporally the definitions account for slow onset disasters, like the impacts of climate change, and sudden disasters with a short window for effective response. Furthermore, the effects may be short-lived to continuing into the unimaginably distant future. Disasters occur over a described geographical impact zone, have clear social impacts, and require a response to limit or mitigate the impact. A definition may or may not include an element of culpability and impact on the environment, ecosystems, and nonhuman species (Kreps 1984, Birkland and Nath 2000, Gephart 2004, Porfiriev 2012). Susan Park in this special issue provides a comprehensive review of the history of the definitional task.

The most recent United Nations Office for Disaster Risk Reduction (UNDRR) definition states disasters result in the 'serious disruption of the functioning of a community or a society at any scale due to hazardous events interacting with conditions of exposure, vulnerability and capacity, leading to one or more of the following: human, material, economic, and environmental losses and impacts' (UNISDR 2017). The 'victim' of the disaster in this definition is a community or society – it is the impact on human activities that is of greatest importance. Human activities, capabilities and flourishing are either directly impacted by the disaster event, or indirectly via 'material, economic, and environmental losses'. A disaster is something that happens *to* people directly or indirectly. Its seriousness then seems to relate to the degree of 'exposure', the people's level of 'vulnerability' or their 'capacity' to withstand and recover from the impacts. The definition betrays a resigned expectation: Recurring human-induced disasters will seriously impact human and nonhuman welfare. While the UNDRR's title indicates a goal to limit disasters we can expect that human actions will continue to trigger events that will be disastrous, disrupting lives, livelihoods and the environment in which people live, work and play. It does not suggest a harm is done or perpetrated upon the environment qua *environment*: that is this definition does not suggest that damage to the environment is intrinsically tragic, it is only of concern insofar as humans or communities or economies may be impacted.

Some point to industrial/technological disasters as failures of the system. For instance, Gephart (2004) suggests, ' [a]ccidents involve unintended and untoward events'. Gephart implies humans are ill-equipped to avert accidents in complex systems, that there is an 'inherent catastrophic potential' (21) within modern technological processes and procedures. He suggests corporate short-termism and lack of incentives to think long-term, unregulated markets and worker dependency on wages (and reluctance to criticize processes) mean that environmental costs are externalized. Disasters are, he says, *inherently political.* There is a tendency for officials to downplay both potential and actual impacts – a feature exposed on the Case. Prevention becomes hard because the overriding political discourse minimizes risk factors. Critics and civil society then have 'difficulty stopping or preventing activity that create[s] environmental disasters because *a priori values, beliefs, and knowledge [give] precedence to the needs of capital* and because critics lack[] access to control settings' (22, italics added). Environmental protection is secondary to political desire and decision-making and to economic objectives because of a particular underlying political philosophic standpoint. Consequently, human-induced disasters are an unsurprising outcome.

Perhaps complex systems are not amenable to control. Do we accept disasters are simply a matter of economic expediency and corporate influence on the political process – a 'failure prone context' (Hellström 1998)? Or should we look deeper into foundational philosophies? Are the 'accidents' that lead to environmental disasters simply 'political accomplishments' (Gephart 2004, p. 22)? My argument is that disasters stem not from complexity alone. They are also a product of the way liberal philosophy and liberal political systems approach complexity. Put more precisely it is philosophy that holds the key to how we contain the unpredictability of complex systems. I will start here with a (very) brief historical survey – of philosophy, politics, colonialism and the settler state of Aotearoa.

Enlightenment, expansion and settlement

The Enlightenment undoubtedly brought great change and new freedoms to the people of Europe. Emerging from a period where knowledge and power were the guarded domains of the elite few, and ignorance and servitude the lot of the multitude, it brought great liberty – the liberties of equality before the law, justice, freedom (including freedom of movement), expansive reorganization of property rights, the turn to 'rational' thought over religious superstition, and representative democracy (albeit selectively male representation). These advances secured individual human wellbeing as central to good governance. People moved from being (mere) subjects to become individualized independent persons. Importantly though these moves established all-that-is-not-human as simple objects outside the spheres of justice and subject to human domination. The nonhuman realm (property), without intrinsic value or comprehensive protection, became framed as 'resources' ripe for exploitation/extraction. The focus on advancement and knowledge also heralded the growth of positivistic scientific and technological discovery. Simultaneously there was the explosion in capitalist enterprise and the expansion of empires to feed industrial machines, to seek territory to absorb the growing populations of Europe, and to acquire property (Stewart-Harawira 2005).

The right to own property became secondary only to the protection and ownership of the self. John Locke legitimized taking lands in the great global colonial land-grab. He claimed the right to ownership of physical property (land) was something one acquired through settlement and 'improvements' wrought through labor. Where someone used the land but did not 'improve' it through effort (as was claimed of non-European Peoples) they had no legitimate claim of ownership (See Locke 1997, and specifically The Second Treatise, Chapter V, ¶ 25 to ¶ 51). By moving onto that land, settling and 'improving' it through clearances (of people, animals, trees, native grasses, native fish), cropping, and grazing activity, colonial theft was legitimated. It was in this Enlightenment period liberal political philosophy formed, a period when persons and their property and possessions united with government to enter into a political community (Locke 1997). The period entrenched a series of exclusionary binaries: human/nonhuman; white/non-white; civilised/wild; man/woman; culture/nature; subject/object; ration/superstition; philosophy/myth; science/folk knowledge/wisdom. These binaries formed the corpus from which settler state politics, institutions and law drew their credibility and legitimacy – and still do. This is the philosophic foundation for the definitions and politics of environmental disasters.

As hinted above the Enlightenment heralded the expansion of colonial empire. In Aotearoa the British colonists came not simply to strip the islands and seas of natural resources, but to settle it with migrants from England, Scotland and Ireland, all under the rule of the British Monarch. But the land was already settled – and settled it was with villages, established horticultural activity, expansive fisheries, linked by nationwide trade routes and comprehensive political and social structures, philosophies, spirituality, education, arts, and medicinal practices. There could be no pretending these were nomadic peoples or that the land was *terra nullius*. Consequently, the British formalized their dominion through 'a' treaty with Māori – the Treaty of Waitangi is the English language version and Te Tiriti o Waitangi the Māori version. I differentiate the versions because they are not exact translations. What Māori signed up for in Te Tiriti is not what the English version says they agreed to (Henare 2007). This is not the place to examine this duplicity (Henare 2007), but rather it heralds two relevancies. The first is that the treaties were repeatedly breached by the settler government from shortly after signing (Henare 2007). Second, since the 1980's the government of Aotearoa has been engaged in programs of negotiation and reparations for the breaches (NZ Government 2020). Included within the parameters of the agreements and in the spirit of good intention, consultation with Māori has become embedded within much legislation.[3] This too is relevant to what follows. Despite the colonial desire for the 'indigenous people to vanish' (Veracini 2015, p. 88), Māori are reclaiming rights as agreed within Te Tiriti. However, that does not mean Māori philosophy, practices, protocols and institutions of environmental and intergenerational justice are given legitimate place or significant standing within the liberal democratic institutions of the state. And from a disaster mitigation perspective, that is short-sighted.

What follows is an exploration of one case study[4] that highlights the way Enlightenment philosophy is embedded within the government, institutions and legal systems of the state and the ways that may be thought of as facilitating disasters. Each aspect is contrasted with Apanui's philosophy into which is woven a precautionary approach that supports intergenerational and environmental justice, and actively mitigates the potential for disaster.

Te Whānau ā Apanui vs the Minister for Energy

In March 2011, with a five-year New Zealand government license in hand, Petrobas[5] maneuvered the seismic survey vessel Orient Explorer into Apanui rohe (territory) in preparation for seismic testing. They came to find oil and gas, and possibly build the world's deepest offshore oil rig. The government and Petrobas were eyeing black gold, twenty-five years of potential riches. Directly challenging the Crown's authority and Petrobas' license Apanui and Greenpeace New Zealand assembled a flotilla of craft to disrupt these endeavours.[6] Mindful of the Deep Water Horizon disaster, Apanui and Greenpeace were seeking to avert a potential disaster. Apanui were fulfilling their obligations to the seas, sea kin, the shoreline, ancestors, the living and future generations. They were asserting mana moana (territorial authority over those seas).[7] Launched from Auckland on 27 March 2011 it arrived in Apanui rohe on 2 April where members of the flotilla were welcomed by iwi (tribal group). After two days much of the contingent returned to Auckland. Those remaining undertook a campaign of direct action, moving into the path of the Orient Explorer, disrupting Petrobas' survey. On 12 April the New Zealand Navy boarded protest vessels and issued notices demanding they remain at least 200 meters from the Orient Explorer. The captain of Apanui's San Pietro was subsequently arrested, after once more maneuvering his boat in front of the survey vessel causing it to alter its course. Still the protest continued. Come 9 May the surveyors left the area – their surveying complete – and the flotilla returned to shore claiming victory. That was not the end of surveying. Petrobas returned and recommenced their survey activity and the fight continued in a new domain through 2011–2012, the courts. Encountering financial difficulties at the end of 2012 Petrobas abandoned their exploration and the government reopened applications for rights to the area (New Zealand Herald 2012). In April 2017 the New Zealand government announced it would ban all *new* offshore oil exploration (although it is important to note there are old permits which continue to stand) – as part of its tilt towards a carbon-free future (Scoop 2017; Ainge Roy 2018) and by mid-April 2020 all the country's offshore exploration for oil and gas had ceased (Scoop 2020) and the last of the exploration vessels exited Apanui's rohe. Apanui's resistance to Petrobas sparked iwi resistance across the nation and may be credited in part with influencing government policy. What is important here is that Apanui had, and have, under their own custom (forged within Māori philosophy and knowledge – Matauranga Māori) an obligation and duty to act to avert potential disaster.

The flotilla was an act of resistance to a process rooted in colonial assumptions about property ownership, environmental values, and liberal philosophy. These specific assumptions and values will be explored here, along with the procedures of state and law. The potential disaster that Apanui and Greenpeace were attempting to avert was an oil spill. If the Raukumara Basin proved to hold sufficient reserves of oil and gas to warrant drilling, the rig would be the deepest offshore oil drilling operation in the world. Furthermore, it would lie within an active seismic zone with the potential to release a magnitude 8.5 earthquake (Scoop 2017), and proximate to Aotearoa's most active volcano, Whakaari/White Island.[8] Superficially this seems foolhardy. However, the responsibility of the Minister is to maximize income from mineral resources. The Minister's responsibilities specifically exclude considerations of environmental impacts of mining operations. This is the responsibility of a separate ministry, who were required to make an environmental impact assessment on application by the company after a license was granted.[9]

Secondly, the Minister is directed to simply consult with iwi before making a final decision. The ministry sent an email to Apanui – and other iwi who were likely to be directly or indirectly affected by the operations. In response Apanui indicated they were dealing with different government department as they prepared submissions on revisions to the Foreshore and Seabed Act and asked for time to complete those before addressing this issue. However, under the terms of the Minister's procedural duties, the obligation is only to 'consult'. There is no stipulation as to the form of consultation. While the law and legislation may seem to be complying with the United Nations Declaration of the Rights of Indigenous Peoples for instance and seeking the prior and informed consent of iwi, in this case it appears simply window-dressing. There was no genuine attempt to accommodate the iwi most likely to be affected by the exploration and drilling,

Apanui, nor to engage on terms that assisted Apanui – for instance, Apanui requested a single governmental contact rather numerous departments and agents – nor to accommodate Apanui concerns. The Minister and ministry and, as we will see, the law are molded from liberal frameworks and fail to find space for different world views and cultural traditions.

As discussed earlier, definitions of disasters shed light on how western culture frames politics, society, individuals, corporates, economic preferences and the environment. An underlying epistemological and ontological foundation is (silently and stealthily) embedded within the state's approach to disasters. This liberal political philosophical base also undergirds the Minister and bureaucracy's proposals to open a gas and oil concession within the Raukumara Basin off-shore from the East Cape of Aotearoa. The 'a priori values, beliefs, and knowledge structures' (Gephart 2004, p. 22) noted earlier, are evident in the submissions presented in the Case, which was Apanui's final recourse to avert exploration and mining in the area. Apanui's objective was clear: to avert a Deepwater Horizon-like disaster.

My suggestion is that the government's values, beliefs and knowledge structures are not unique to the politics of Aotearoa but rather that they underpin politics, legislation and regulation in the Anglo-European-American spheres and seep into international regulation. Simultaneously, that these philosophic underpinnings are inadequate for environmental, multispecies and long-term community justice and for protection from disasters emerges in stark contrast with the already always anticipatory and inclusive philosophical foundations of Apanui's obligations and duties to the environment and intergenerational justice. How did the parties find themselves at loggerheads on beaches, at sea and in the court?

In 2002, New Zealand's government[10] decided it wanted to increase revenues from hydrocarbon deposits. Specifically, they identified that '[o]f particular relevance to the management of the Crown mineral estate is one of the economic objectives of MPP [Minerals Programme for Petroleum] that New Zealanders, "Derive considerable value from our natural advantages in terms of resources, climate, human capital, infrastructure and sense of community"' (RNR13 §8:3). Oil already contributed 1% ($850 million/year) to Aotearoa's GDP (RNR13 §15:4), however with existing fields about to close the government sought new mines to revive non-tax revenues (Winter 2018). Responding to a government directive the Ministry of Economic Development crafted a revised policy objective: to find new oil and gas reserves that showed potential for mining (RNR13 §2-§3). With their presumptive claim of sovereignty over mineral resources and coastal waters and without a touch of irony, the Crown set about exploring for offshore oil within Apanui rohe, decimating the very 'sense of community' they describe as integral to the MPP (Winter 2018).

The Department's claim that '[p]romoting exploration and discovery of new oil and gas fields provides for the nation to use to best advantage its natural geological resources' (RNR13 §19:5) reflects their epistemological and ontological certainties. This one phrase tells us: that which is natural (nonhuman) is an object, here phrased as a resource; that the resource is something separate to the context in which it is embedded; that the thing of importance is 'best advantage' which might be interpreted as an economic objective. This is Locke transposed into the twenty-first century. The objective was that 'the Crown, as owner of the petroleum resource, should obtain a guaranteed minimum royalty payment from the extraction of its petroleum' (RNR13 Appendix 1:9). The government's resolute desire was to extract maximum financial value from Aotearoa's petroleum resources over a twenty-five-year period.

The Petroleum Act (1937) (PA), §3 declares petroleum is the 'property of the Crown.' While Māori have repeatedly contested this ownership claim (based on Article 2 of Te Tiriti which preserves an interest in their taonga[11] (RNR 28)) the government has consistently asserted an exclusive Crown property right over minerals. The title to land, sea, lake or river bed may be held in individual or communal ownership, however, the mineral resources under the surface may not – they belong to the Crown. This repeated declarative ownership, based on the principles of colonization, isolates mineral resources from their context. Minerals are an object which may be, should be, extracted for wealth and becoming property. The intricate nexus of mineral, rock, soil, waters, air, currents, and the allied wellbeing of living things, is peripheral to this understanding of ownership and property rights. The government's worldview underpins a fragmented ownership framework, dissociating the minerals from their wider environmental and community context (Winter 2018). The environment is divided into discrete spaces and temporalities (Bond 2013). Neither environmental diversity nor ecosystem health are understood here as wealth.

The government's description of the environment – the block offer – is reductionist, mechanical and objective. In Figure 1 below, a series of points of latitude and longitude, degrees and minutes, and compass points are used to describe 'The Land'. This is the government's description. Using the tools of rational science and the language of oceanic exploration anyone can locate The Land with great accuracy – with the right kind of education. This description, and its representation on a map as two discrete rectangles placed out to sea from the East Cape shoreline, are 'rational' descriptors. The ocean floor and the geomorphic substrate are divorced from the currents in the sea, the flows of brine that slide through the rectangular boundaries like ghosts. Like ghosts too are the whales, the fish that glide with, across and against the currents. There is nothing of the sublime in these descriptions – none of the feelings of wonder that well in the human heart in an encounter at the edges

Latitude		Longitude	
36° 30' 00.0000" S	and	177° 30' 00" E	then east to
36° 30' 00.0000" S	and	178° 10' 00" E	then south to
36° 40' 00.0000" S	and	178° 10' 00" E	then east to
36° 40' 00.0000" S	and	178° 30' 00" E	then south to
36° 50' 00.0000" S	and	178° 50' 00" E	then east to
36° 50' 00.0000" S	and	178° 50' 00" E	then south to
37° 00' 00.0000" S	and	178° 50' 00" E	then west to
37° 30' 00.0000" S	and	179° 00' 00" E	then north to
37° 30' 00.0000" S	and	179° 00' 00" E	then west to
37° 30' 00.0000" S	and	177° 30' 00" E	then north to

Figure 1. The Land (replicated from RNR 14, GAB4, Petroleum Exploration Permit 52707, p3 & also reproduced in Winter 2018).

of our ken. There is nothing here that describes any ancestral connection with this place, no sense that these boxes of flowing water and slowly morphing seafloor have provided sustenance to generations of human and nonhuman beings. There is no sense of nurture nor responsibility nor relationships nor care.

Not only the tool of navigators, these descriptions of longitude and latitude are also the tools of the law. They describe property – that which is owned. That which may be used (and abused) largely (with some caveats) at the discretion of the owner. The descriptors and the law strip from The Land its entanglements, context and flows. Significantly it is divorced from the active seismic and volcanic presences just outside these bound artifices. Instead it becomes mono-faceted and fungible. Something to be traded for something of equal value. The value to the government is the potential of twenty-five years of royalty revenue from the eventual concession holder. And the expectation of even more value – the profit the concession holder expects to generate over the same 25 years. Profits – the surplus to the expenditure of capital and labor (with minimal obligations to environmental recovery or regeneration). Again we hear Locke echo. The treasure/taonga here is condensed to an abstract representation of ownership and labor, the dollar. The value of the sublime, the critters, the currents, the history of the place bear no reckoning in this calculation and descriptors. Those become the 'other' of Enlightenment-informed dichotomization: human/nonhuman; culture/nature; civilization/wilderness; rationality/superstition.

The scientific language of the Minister and his bureaucrats defines and assigns boundaries. Apanui's description animates The Land. The Land is a flow of relationships, of spiritual connection, intergenerational responsibility, care, and reciprocal obligations. From The Land come identity, gifts and kai moana (sea food). Mana moana comes from fulfilling responsibilities to the rohe.

When environmental protection is primary[12]

While the Minister submitted forty plus affidavits and pieces of documentary evidence to the Court, Apanui's evidence rested on one 21-page document – the Affidavit of Te Kou Rikorangi Gage. It starts with a statement of Gage's background and positions – including his comprehensive knowledge of tikanga and the history of his people, and his many appointments to national boards and governing bodies. He is bilingual and bicultural (Gage 2012, §1). The purpose of the affidavit is to 'clearly state my tribe's understanding and views concerning its historical and current relationship and responsibilities in relation to our tribal lands and sea area' (Gage 2012, §9).

Gage attempts to create a communicative bridge across the ontological divide between the Crown and Apanui. He states he wants to establish a 'conceptual framework as a type of "toolkit" that hopefully will assist in interpreting and appreciating our worldview, our tikanga (laws, customs and practices) and more importantly the deep spiritual nature of our interconnectedness with Te Ao Turoa (the natural world; our lands and territories)' (Gage 2012, §11(a): 1). Understanding that conceptual framework is critical to understanding Apanui's relationship to land and sea. Importantly, it indicates 'why [they] would never cede [their] mana or [their] tribal inheritance over to the Crown' and stresses iwi concerns over activities the 'may threaten the wellbeing of our mokopuna, and also the web of life and systems that our natural world provides' (Gage 2012, §13(a)&(b): 2)'. Mokopuna are future generations.

In contrast to the dichotomized, mechanical and economic descriptions from the Crown, Gage clearly establishes for the Court Apanui's *relational* conceptualization of the world from which obligations and duties arise. Māori may now own title to lands as part of a post-colonial heritage, however iwi territory is more than a material asset, it is a source of identity. The whakapapa of the land and people is intertwined, it is 'the means by which the interrelatedness of all things ... is identified to our consciousness and our spirit. It relates us to the past and to our future, and to our mokopuna (descendants)' (Gage 2012, §46: 7). A genealogy of all things, spiritual, human and nonhuman, built up in layers from the beginnings of time, through ancestral links to the present the whakapapa guides the iwi's relations and relationships with the territories over which they hold mana and to which they hold obligations of kaitiakitanga. Whakapapa is the foundation of the relationship: from it flows an understanding of intrinsic value in and interconnectivity between everything. When Apanui ask themselves the question, 'How must I relate?', the domain of relationships is all things – individually and as a collective whole. The whole and the things that make it up all bear value.[13]

Tapu, mana, and mauri describe intrinsic spiritual elements of being – in all things, human and non-human (Winter 2019). Translating them is difficult and translations vary by iwi, and author. Tapu, mana, and mauri encompass concepts of respectively, the holy and potential to be, dignity and authority, and life force. Importantly, these are intrinsic to all being and consequently, '[d]istinctions between inanimate and animate objects are ... blurred ... Nothing is lifeless' (Durie 1998, p. 23). They connect the spiritual realm with the temporal, linked through whakapapa and convey the message that 'we are part of an interconnected and interrelated whole' (Gage 2012, §18: 3). Through them, the relationships between people, fish and birds, plants, seas, rivers and land are created in a web of eternal respect and mutuality.

Nonhuman are kin, a part of the layered stratum of interconnections to be nurtured not exploited. 'The idea of what is referred to ... as "resource management" does not sit well with us, as it stems from a framework that sees these resources as things to be used for the benefit of human needs. Our framework sees these things as beings possessing their own tapu and their own mauri and who manage themselves quite happily without the intervention of man' (Gage 2012, §31: 5).

Apanui have an ethic of care and reject 'resource management'. They also harvest from the sea. Gage notes the possibility that offshore drilling operations will pollute and undercut that potential for harvest is a major concern for Apanui. Particularly he notes two groups for whom kai moana (sea food) is of ongoing importance. Firstly, for the poorer members of the iwi for whom gathering kai moana is necessary to supplement their diet. And secondly, to the iwi as a whole whose mana (prestige) is closely tied to its ability to host visitors and for whom providing feasts of kai moana is historically and continuously integral to that mana. So not only is protection of the seas important spiritually, ethically and morally, so too the iwi has practical, internal political duties and obligations as kaitiaki (guardians) and hosts.

Intrinsic to these material responsibilities are responsibilities to the sea, sea floor, sea shore, and the living plants and creatures that live there. Operating from an epistemology of interconnection, and reciprocity the nonhuman is intricately connected to human, part of a web of spiritual and material kinship relations. Significant toanga species, whales and penguins are known to be disturbed and harmed by the sorts of seismic testing required for locating the oil and gas reserves. Apanui have an obligation to protect these species from harms. This is both a moral and political duty – one attached to mana moana, the claim to territoriality, and the respect that comes with proper guardianship of the area and all within it. Resistance to this proposal is founded in both short-term known events (the seismic testing) and long-term unknowns (oil spills).

Apanui iwi management practices aim to maintain and protect biodiversity on their lands and seas, to enable gathering of kai moana, critical to the iwi's hospitality to visitors and provide 'backstop' food sources to 'economically depressed' iwi members (Gage 2012). This is a relationship of sustenance and longevity and includes management practice of rahui, a temporary or permanent ban on entering or taking resources from a described area. Gathering sea food is conducted always with respect for the 'spiritual powers and vitality of mauri to which the productivity of the food gathering areas is inextricably linked' (Gage 2012, §35: 5).

The marine environment within their rohe is part of the iwi's taonga. Discrete separations between human and nonhuman are not emphasised in this worldview, and neither are there separations between seafloor and sea. The seafloor, sea, foreshore and the creatures within form a continuous whole as part of the iwi's rohe. Care for the sea and the creatures place obligations on Apanui as kaitiaki arising from mana moana. Everything in this Māori worldview is interconnected and overlapping. The separate elements cannot be corralled by boundaries drawn on maps.

This contrasts sharply with the Crown's description where the foreshore, sea, sea floor and oil deposits below are disjointed: and marine life forms are rendered immaterial ghosts. For Apanui everything is immutably interconnected not a measured resource-bearing property, but beloved places upon which iwi identity is contingent.

During the marine blockade Apanui expressed concern at siting the deepest off-shore drilling operation in the world sitting beside an active volcanic and seismic zone (matters not addressed by the Minister). Apanui stresses the risk of an accident polluting their territory. A spill could not remain within the discretely bounded concession blocks, but rather spread through the waters, despoiling water, sea bed, and shore and endangering life and livelihoods. Such a spill would hinder their role as kaitiaki – rendering them powerless to protect the territory and taonga within it, unable to fulfil political, cultural and moral duties and obligations *as Māori*. The issues here are not only philosophical, but also practical. Aware of the risks with these operations, Apanui express concern their particular obligation, resting in specific duties to ancestors, to pass the environment to future generations in a better condition that they received it is threatened. While risk management requirements may be enforced by various legislation Apanui are clear the damages from an oil-spill are very long term. While commitments under the law to resuscitate the environment may be forced on the miner, nothing could return the area to a pristine state.

Gage stresses social, spiritual, material/economic and environmental values pervade all Apanui actions and decisions. That is, every decision in the material realm is screened through spiritual, social and environmental lenses. Every decision has an intergenerational dimension and is conducted in 'partnership' with the environment. As guardians, Apanui are responsible to ancestors for providing for future generations. They are responsible to the mauri, tapu and mana of the environment itself to ensure more than protection, they must improve on what they receive. There are no silos in this ontology, there is rather an interlocked, spiral flow of non-living to living to non-living, ancestors to present to future generations, land to sea to human to non-human, spiritual to political to material to ethical. Immersive connections and functioning guide relationships, obligations and duties. Reciprocal responsibilities flow between all elements of these relationships.

The Apanui relationship to territory described by Gage is the antithesis of the Crown's; '[the] terms of our relationships with the environment, ... are governed by a tikanga (laws) of deep respect and spiritual bond' (Gage 2012, §18: 3). They interconnect spiritual and temporal realms, the continuous overlapping places, described through genealogical relationships in which people and the environment are part of one family, 'in which we see ourselves as but a 'tiana' or younger relative (Gage 2012, §31: 5).

Kaitiakitanga frames a set of obligations and duties that are integral to being Apanui a necessary expression of identity, culture, political influence and determination.

And yet the High Court found in the Minister of Energy's favor – he had followed the processes required by law.

Discussion

I have looked at the influence of liberal political philosophy in two domains; the UNDDR definition of disasters, and the New Zealand government's decision to open deep sea oil deposits to exploration (and potential mining) adjacent to an active volcano and seismic zone. Apanui's philosophical constructs were offered as a counterpoint, an alternative to the liberal tradition. I argue that the philosophical roots and intentions of liberalism go largely unexamined in policy formation and institutional arrangements. Stripped of philosophic deliberation the excesses of human desire for mineral wealth go un- or less-checked. While the government and courts can entertain discussions of process and procedure they are unable to grapple with and weigh the philosophical. So my argument has been that this was not a matter of testimonial injustice (Fricker 2007) – that is an injustice rooted in lack of belief in credibility of the witness. Gage's credibility is evident in his extensive experience within and secondment to mainstream organizations and committees. What we witness is a form of epistemological injustice (Fricker 2007) in which the very structure of the knowledge presented could not be 'heard' by the Court. The Court could only 'hear' a corpus of rules that legitimate extraction at the expense of the environment and Māori communities.

Likewise, my argument is that those who are formulating laws and regulation – national and international – to protect the environment and communities from (inevitable) disasters devise such laws and regulations from within a given philosophic frame – a framing designed during the Enlightenment that separates the human realm from the nonhuman and spiritual realms, and structures (or legitimates) relationships based in human domination of the nonhuman. Within such a structure to monetize the nonhuman realm is both a foregone inevitability and an ultimate goal. This philosophic frame is spotlighted by the MPP which overtly states that the Minister's primary objective must be to extract 'value' from the nation's mineral reserves. This is not to say that the government has not established other laws and mechanisms as environmental safeguards, they have. The point is they are triggered *after* the mining licenses are granted and are not integrated into the initial planning for resource extraction. The assumption seems in this case to have been that protections can be devised that both allow the mining to continue and afford some protection to the environment and communities on the shore. That is, the idea of disaster or the potential for disaster is a secondary consideration. Furthermore, these assumptions are integral to the Minister's responsibilities and the legislative frameworks of the country.

Like the ocean blockade, Apanui failed in its High Court bid to halt explorations. They based their challenge on a claim the Minister had not followed the correct procedure when granting Petrobas its license. However, the court found that the Minister *had* fulfilled his obligations under the legislation. It could not examine his compliance with the processes required under Apanui philosophy and law, for those are not legislated. The Minister's actions were strictly prescribed: he must optimize revenue potential from the nation's oil and gas reserves, and he must consult Māori in territories within and in close proximity to the sites. However, the grounds for Māori objection is tightly restricted. That all territory and everything within it is sacred escapes the Minister, Courts and boundaries of legislation.

That the spiritual and the nonhuman realms escape those boundaries is a function of and inheritance from, a philosophy forged in the Enlightenment. In it human endeavor and desire, and human beings are elevated. They are positioned as external and superior to nonhuman beings and the environment and beyond the complex entanglements of the biosphere. The 'rational' human is seen as the desirable endpoint of evolution against which all else is measured and found inferior, and thus of no moral consequence. 'Rational' humans are bent on scientific discovery, technological advance, environmental mastery, and capitalist enterprise. The land and seas are to be conquered and improved, taken and profited from. Humans can be expected to 'master' disaster events (Bond 2013, p. 697). It is these tenets that make disasters no surprise. As long as human enterprise is separate from and superior to nonhuman enterprise and existence it is privileged over protection of the latter. The illusion of control leads to hubris. A hubris that means the early definitions of disasters might omit mention of the environment, and later definitions normalize disaster (see Gephart 2004). Human-induced environmental disasters are, as it were, a normal part of doing business, and externality of some significance but less significance than the drive to extract resources and profits. Oversight and restraints are carved off to agencies, and seen as an impediment to the more important business of business.

That environmental protection is seen as an impediment is critical at this juncture. Despite the ongoing loss of species, habitats, the damage to the environment, society and economies of climate change the philosophic foundations states operate from are so ingrained nonhuman continue to be seen as other and either valueless or as resources for extraction. Through this window disasters must be expected and accepted in the pursuit of profit and growth. Governments measure their success or failure by growth in annual GNP/GDP not by success or failure to protect the environment – either for the living or future generations. We do not receive daily media reports of daily/annual/decadal extinctions, hectares denuded, soil-inches lost, acres of trees removed, or cubits of waters polluted – they are displaced by daily fluctuations of global stock markets. Moreover, the deaths, illnesses, displacements and losses of humans from disasters fades quickly from memory – until the next event at which point they become disembodied data points of comparison. This is entirely consistent with a philosophy that elevates capitalist enterprise. And in such an environment, disasters are no surprise.

I have provided a philosophical contrast, underpinned by kaitiakitanga which continues to be rehearsed and reinforced by Māori. Within their communities 'buy-in' and commitment to long-term sustainability and environmental protections are easily established and maintained by Apanui and other iwi. Robust and pre-emptive disaster mitigation is an integral and necessary element of their philosophy and practice of intergenerational environmental justice. In the context, then, of human-induced environmental disaster Apanui has the *more developed* philosophy: it recognizes the interconnections, flows, relationships and complexities of the seascape, land and earth system. It is impossible within this philosophy to separate out the parts, to describe sea as a mere rectangular overlay on a two-dimensional map or grid references in a chart. The lives and entanglements of human, fish, sea birds, plants, shore, seafloor, ocean, currents, winds, waves are all always party to disaster mitigation planning and practices. Disaster then is something truly unexpected and surprising.

Notes

1. I am a New Zealander of Anglo-Celtic-Ngāti Kahungunu descent, and a settler in Australia living and working on the unceded lands of the Gadigal People of the Eora Nation.
2. In accordance with decolonial practice Māori words will not be italicised. Wherever they are first used they will be contextually or parenthetically defined or explained.
3. See for instance the Resource Management Act (1991) and the Marine and Coastal Area (Takutai Moana) Act 2011.
4. The source materials are affidavits and documents accompanying testimonies in CIV-2011-485-1897 [2012] NZHC 1422, Greenpeace of New Zealand Incorporated and Te Runanga o Te Whanau-a-Apanui vs. The Minister for Energy and Resource and Petrobas International Braspetro BV heard in 2011–2012.
5. Petrobas International Braspetro BV is a subsidiary of Petrobas a Brazilian energy company. On 1 June 2010 Petrobas and the New Zealand government formalized an agreement that granted Petrobas 100% of rights to Block 2. In return Petrobas agreed to undertake a minimum work program, including two stages of seismic testing followed by the drilling of one exploratory well. On 5 December 2012, Petrobas announced it was withdrawing from New Zealand and handed back its license a move prompted by their global financial losses (New Zealand Herald 2012).
6. This flotilla may be seen as a continuation of an earlier campaign of beach protests on 27 June 2010 which were followed by a series of bonfire protests on the beaches of the East Cape.
7. Mana moana comes with a set of reciprocal obligations and responsibilities to maintain and care for the territory and to pass it on to the next generation in a better condition. This is the heart of the obligations of kaitiakitanga.
8. Whakaari/White Island was the site of a violent eruption in December 2019, a disaster which killed 21 tourists and injured and maimed 26 more while they were exploring its crater – part of a steady stream of visitors taken to the island daily.

9. I have not engaged with the procedures undertaken by the Department of Environment. While they too may make an interesting case-study, Te Whanau-a-Apanui did not take that department to court and I have restricted my examination to this particular legal battle.
10. Which I will refer to as 'the government' or 'the Crown'.
11. Taonga: treasures, valuable possessions, property, goods.
12. Much of the following section is drawn from my PhD thesis - Winter (2018).
13. See Jeremy Bendik-Keymer for an examination of the relational dimensions of morality. Applying his account of evil to this Māori worldview would then mean an 'unacceptable rending of human relationships' to all things, 'forests and oceans, fish and fowl, the rivers and the soil and between people and the elements' (in Mason Durie's words). That is, if evil results from destruction of the intrinsically valuable, it results from the pointless, or unjustifiable destruction of any part of the natural world (Bendik-Keymer 2013).

Acknowledgments

My sincere thanks to Susan Park for encouraging me to write this and for her insightful review of multiple drafts. The case study research was undertaken as part of my PhD thesis: thanks are due David Schlosberg for hours of discussion, critique and encouragement. Thanks, too, to three anonymous reviewers for their thoughtful suggestions and positive reception of this endeavor.

Disclosure statement

No potential conflict of interest was reported by the author(s).

ORCID

Christine J. Winter http://orcid.org/0000-0002-5427-9196

References

Ainge Roy, E., 2018. New Zealand bans all new offshore oil exploration as part of 'carbon-neutral future'. *The Guardian*, 11 Apr 2017. Available from: https://www.theguardian.com/world/2018/apr/12/new-zealand-bans-all-new-offshore-oil-exploration-as-part-of-carbon-neutral-future

Barnett, M. and Finnemore, M., 2005. The power of liberal international organizations. *In*: M. Barnett and R. Duvall, eds. *Power in global governance*. Cambridge, New York, Melbourne, Madrid, Cape Town, São Paulo: Cambridge University Press, 161–184

Bendik-Keymer, J. (2013). The Moral and the Ethical. *In*: V. Gluchman, ed. *Morality*. Amsterdam: Rodopi, 11–24.

Birkland, T.A. and Nath, R., 2000. Business and political dimensions in disaster management. *Journal of Public Policy*, 20 (3), 275–303. doi:10.1017/S0143814X00000854

Bond, D., 2013. Governing disaster: the political life of the environment during the BP oil spill. *Cultural Anthropology*, 28 (4), 694–715. doi:10.1111/cuan.12033

Durie, M., 1998. *Te Mana, Te Kawanatanga*. Auckland: Oxford University Press.

Fricker, M., 2007. *Epistemic injustice: power and the ethics of knowing*. Milton Keyes: Oxford University Press.

Gage, T.K.R., 2012. *Affidavit of Te Kou Rikirangi Gage* (J.J. Gendall, Ed.) (June 22, 2012). Wellington: High Court of New Zealand.

Gephart, R.P., 2004. Normal risk. *Organization & Environment*, 17 (1), 20–26. March. doi:10.1177/1086026603262030

Hellström, T., 1998. The Exxon Valdez disaster: environmental safety implementation as an incrementalist mode. *Social Science Information*, 37 (2), 361–380. doi:10.1177/053901898037002007

Henare, A., 2007. Toanga Māori. *In*: A. Henare, M. Holbraad, and S. Wastell, eds. *Thinking through things*. London and New York: Routledge, 47–67.

Kreps, G.A., 1984. Sociological inquiry and disaster research. *Annual Review of Sociology*, 10, 309–330. doi:10.1146/annurev.so.10.080184.001521

Locke, J., 1997. *John Locke* (P. Laslett, Ed.). Cambridge, Melbourne, New York: Cambridge University Press.

New Zealand Government, 2020. *Settling historical Treaty of Waitangi claims*. Available from: https://www.govt.nz/browse/history-culture-and-heritage/treaty-of-waitangi-claims/settling-historical-treaty-of-waitangi-claims/

New Zealand Herald, 2012. *Brazilian oil giant Petrobras dumps NZ exploration permits*. nzherald.co.nz. Available from: https://www.nzherald.co.nz/business/news/article.cfm?c_id=3&objectid=10851863

Porfiriev, B., 2012. Reflections: environmental hazards & disasters: research perspectives. *Environmental Hazards*, 11 (4), 324–327. doi:10.1080/17477891.2012.736216

Scoop, 2017. Multinational Probe of Hikurangi Subduction Zone. *Scoop Politics*, 16 Oct. Available from: https://www.scoop.co.nz/stories/SC1710/S00035/multinational-probe-of-hikurangi-subduction-zone.htm

Scoop, 2020. OMV Exit Signals End of Offshore Oil Exploration in NZ. *Scoop Politics*, 10 Apr. Available from: https://www.scoop.co.nz/stories/PO2004/S00095/omv-exit-signals-end-of-offshore-oil-exploration-in-nz.htm

Stewart-Harawira, M., 2005. *The new imperial order*. Wellington: Huia Publishers.

Tallberg, J., *et al.*, 2020. Why international organizations commit to liberal norms. *International Studies Quarterly*, 64, 626–640. ISSN 1468-2478. doi:10.1093/isq/sqaa046

UNISDR, 2017. *Terminology, Geneva, Switzerland, United Nations International strategy for disaster reduction*. Available from: https://www.undrr.org/terminology/disaster

Veracini, L., 2015. *The settler colonial present*. Bassingstoke UK, New York: Palgrave.

Winter, C.J., 2018. *The paralysis of intergenerational justice: decolonising entangled futures*. Thesis (PhD). University of Sydney. Available from: http://hdl.handle.net/2123/18009;oai:ses.library.usyd.edu.au:2123/18009

Winter, C.J., 2019. Decolonising dignity for inclusive democracy. *Environmental Values*, 28 (1), 9–30. doi:10.3197/096327119X15445433913550

Winter, C.J., 2020. Does time colonise intergenerational environmental justice theory? *Environmental Politics*, 29 (2), 278–296. doi:10.1080/09644016.2019.1569745

Extractive industry disasters and community responses: a typology of vulnerable subjects

Teresa Kramarz

ABSTRACT
There is widespread recognition that environmental disasters disproportionately affect the poor in developing countries, yet certain kinds of violence also strip white, middle-class communities in industrialized countries of vital resources to respond. Social vulnerability as a mutually constitutive relationship between distinct types of shocks and capacities to respond requires further research. I explore how the temporal dimension of shocks creates different types of structural risks, and articulate how these interact with the agency of affected communities. The first section deals with attributes of different environmental shocks, the second examines the agency of individuals and communities, the third analyzes the relationship between extractive industry violence and capacities for response. This generates a typology that enables us to chart the creation of vulnerable subjects. This framework also helps us develop insights into the underlying dimensions of social vulnerability and makes it possible to better classify cases of vulnerable populations.

Introduction

Rapid onset environmental disasters are spectacles that concentrate – at least for a time – public attention on a dire event. However, media coverage that focuses on such events, from oil spills to mining catastrophes, masks the causal chain of issues underpinning the devastating impact of a disaster and privileges the visibility of tipping points. These causal chains are also discounted in everyday public discourse (Ribot 2011). For example, the health and environmental damage that comes from living next to a petrochemical compound where heavy metals permeate the water, air, and soil of poor communities, and where garbage dumps, lack of sewage, and access to clean water create chronic diseases, is part of the social vulnerability of everyday life for many communities that lies largely dormant when public debate and the scholarly literature on building resilient societies eschews addressing root causes.

Like the COVID-19 pandemic, environmental disasters have been treated as singular, catastrophic events that capture everyone's attention. As Park examines in the introduction of this symposium, disasters have cascading effects, yet they are generally considered isolated, linear events. However, the full force of disasters is unleashed by underlying processes of poverty and inequality. As I discuss below, the environmental resilience literature should similarly expand in this direction.

In this article, I focus on environmental disasters caused by mining industries (including those that extract fossil fuels). It deploys a political ecology lens to conceptualize such disasters in broad terms that includes not only the direct physical harm immediately sustained by a rapid onset shock like an oil spill but also by the underlying structural conditions that produce slow and ongoing harms such as living next to a fracking pad, along a pipeline or next to an oil refinery (Nevins 2003). I rely on the concept of social vulnerability to highlight that the state of being susceptible to shocks is a 'socially-constructed phenomenon influenced by institutional and economic dynamics' (Adger and Kelly 1999, p. 253). It is a product of social and place-based inequalities (Cutter *et al.* 2003). Conceptually, social vulnerability is associated with exposures to stress that disrupt livelihoods, particularly of already marginalized communities. It is a concept that links the impact of social inequalities to environmental disasters (Kuhlicke *et al.* 2011).

I intentionally eschew references to resilience, which is often paired with vulnerability in the literature on environmental disasters, because of its emphasis on individualizing risk and top-down technocratic prescriptions for 'bouncing back' that underexamine questions of power and inequality that engender risk (Adger 1999, Ziervogel *et al.* 2017, Chandler 2019). As Popke *et al.* (2016, p. 74) argue, 'resilience approaches are problematic, because they risk eliding important sources of inequality around climate change vulnerability in favor of a systems perspective that emphasizes the creation of responsible, resilient subjects.' Furthermore, Brown and Westaway (2011) claim that the main policy tools used to build capacity and resilience underexamine the role of agency among individuals and families. Many models focus instead on GDP as 'one or the only determinant of adaptive capacity and adaptation' (Grothmann and Patt 2005, p. 206). However, regardless of economic wealth, communities have no choice but to be resilient agents when confronted with debilitating shocks to their livelihoods. Paradoxically, polluting industries and unresponsive governments can rely on the inescapable need of populations to adapt to any shock and continue harmful practices, thereby making resilience a problem rather than a solution (Bohland *et al.* 2018).

Although there is widespread recognition that environmental disasters disproportionately affect the poor in developing countries, this article provides a framework to investigate why certain kinds of violence also strip white, middle-class communities in industrialized countries of vital resources to respond. For example, in this symposium, Gemma Viney documents how an oil seam gas project in New South Wales, Australia depleted white farming communities of key psychosocial resources that had historically enfranchised them. The mutually constitutive relationship between types of shocks experienced and capacities to respond requires further research. I explore how the temporal dimension of shocks creates different types of risks, and articulate how these interact with the endogenous capacities of affected individuals and communities. Investigating why and how social vulnerability profiles differ is not only analytically important but vital to developing more effective policies (Romero-Lankao *et al.* 2016). As Ribot (2013) has argued, vulnerabilities do not just fall from the sky and subjects of vulnerability do not randomly emerge on a scene, but are constructed through social, political and economic institutions and their interactions. Therefore policies to address vulnerabilities need to consider not only who is more or less vulnerable, but in what ways and why?

A first point of departure is the assumption that vulnerabilities are not all created equally. To disentangle the nature of social vulnerability, I focus on environmental shocks and responses that are deeply rooted in historical, cultural, social and economic processes (Cutter 1996, Cutter *et al.* 2003). This builds on previous research tracing historically conditioned processes of vulnerability to hunger and famine and the layering effects of multiple threats including socio economic status, education, gender, race, age, and disabilities (Watts and Bohle 1993, Blaikie *et al.* 1994, Collins 2009, Peek and Stough 2010, Hewitt 2019). Since these processes are historically and geographically contingent, and because extractive industries create localized socio-environmental costs with dispersed economic benefits, the article focuses specifically on community vulnerabilities. Although there are many conceptions of community, this article deals with those normally referred to as place-based or grounded in terms of a geographical location (Barrett 2015, Räsänen *et al.* 2020). Second, I investigate how the external and internal dimensions of vulnerability interact in these communities. Vulnerability is Janus faced, one dimension refers us back to conditions that created it and the other to conditions that risk perpetuating it (Chambers 1989, Bohle 2001, Marino and Faas 2020).

To unpack these dual dimensions the article proceeds in three steps. The first deals with the structural conditions that shape environmental shocks. Traditionally, the disasters literature treats these shocks somewhat generically because scholars consider that 'for most of the risks associated with natural hazards, there is little or no opportunity to reduce the hazard' (Coburn *et al.* 1991, p. 26 as cited in Dilley and Boudreau 2001). I characterize shocks as fast versus slow types of violence to examine different structural attributes and impacts on communities. This distinction also builds on critiques of the resilience scholarship that considers adaptation to extreme and fast events and treats them as shocks that are disconnected from the causal chain of conditions brought on by other kinds of violence, such as that which unfolds slowly and over time (Ribot 2011, 2014, Nixon 2011). Yet, these prior conditions are what have made some communities disproportionately vulnerable in the first place – draining them of their capacities to respond – which requires deeper analytical treatment. As defined and detailed below, the concept of extractive violence in this article encompasses agency and intentionality. It refers to human actions that cause harm, through purpose or neglect, on other human and non human subjects (Shapiro and McNeish 2021).

The second section shifts its focus to affected communities and their agency. It examines the means these have to respond to environmental violence by disaggregating different types of capacities at the individual and community level. The kinds of capacities communities enjoy at a particular space and time are a key variable in evaluating the creation of vulnerable subjects. This section leverages previous research on capacities, entitlements and enfranchisement to articulate different dimensions of community responses.

The third section develops a conceptual typology of vulnerable subjects to explore key dimensions at the intersection of structure and agency. The typology examines the relationship between types of shocks communities experience and their capacities to respond to them. Investigating the politics of differentiated threats and responses can better explain why some disasters are bound to continue to repeat themselves and spiral into ever more devastating events (Branch 2018).

Beyond generic shocks

Here I begin by examining the concept of violence from narrow to broader interdisciplinary understandings that capture the different and historically contingent experiences of affected communities. Violence as a concept and policy problem has most often been treated within disciplinary silos. For example, Krause (2009) identifies politically motivated violence such as war and terrorism as the traditional domain of political science and international relations; economically motivated violence such as organized crime as that of criminology; socially conditioned violence such as riots or ritualistic acts as that of sociology and anthropology; and interpersonal violence including individual pathological, domestic, and gender-based forms as the traditional domain of psychology, public health and criminology.

A more integrated approach to the study of violence has taken place within the folds of various disciplines. In 1969, the sociologist and pioneer of the field of peace and conflict studies, Johan Galtung, significantly expanded interdisciplinary scholarship on violence to include previously neglected dimensions. Among these, Galtung highlighted structural, temporal, and indirect effects. These are important contributions because they enabled recasting the experience of those people and places whose self-realization has been thwarted by avoidable and harmful assaults (Galtung 1969). This broadened academic and public debates on violence.

In 2001, Peluso and Watts's book *Violent Environments* deployed a political ecology approach to delve more deeply into the historically contingent and site-specific contexts that create proclivities for the type of slow violence that Nixon (2011) would later expand. Rejecting the post-Cold War era thesis that resource scarcity leads to conflict, and critiquing it as overly simplistic, Peluso and Watts argued that differentiated patterns of entitlements to resources shapes a kind of political economy of violence (Peluso and Watts 2001).

Nixon's 2011 book *Slow Violence and the Environmentalism of the Poor* leveraged an environmental humanities perspective to focus more deeply on the delayed effects, invisibility and politics of representation that characterize the kind of environmental violence experienced by those locked into structural relationships of dispossession (Nixon 2011). Without this kind of lens, what do we call and how do we communicate acts of expropriation by the state or an oil company that takes land, limits access to resources, or imperils other sources of livelihoods from the poor in the name of development or security? It is in such contexts that Branch (2018) points out how dominant discourses around climate disasters 'occlude the forms of environmental violence and experience of climate change among disaster-affected communities' (306).

Nixon deals with large scale and long-term conflicts, that disperse responsibility and defer accountability from perpetrators. Slow violence, against the poor and the environment, is defined as 'a violence that occurs gradually and out of sight, a violence of delayed destruction that is dispersed across time and space, an attritional violence that is typically not viewed as violence at all. Violence is customarily conceived as an event or action that is immediate in time, explosive and spectacular in space, and as erupting into instant sensational visibility' (Nixon 2011, p. 2).

In summary, a broad conceptualization of violence has been defined as 'an action or a process that appears in visible and unseen forms against humans, nature, and the sustainability of their relations' (Navas *et al.* 2018, p. 650). Three dimensions of slow violence are particularly important in analyzing the creation of vulnerable subjects. These are the unequal distribution of impacts, heterogeneity of marginalized groups, and the invisibility of the struggle. I turn to a discussion of these next.

Unequal distribution of harms

The phenomenon of slow violence does not occur anywhere indiscriminately, just as climate change does not affect all communities in the same way. It is unevenly distributed and there is a pattern in the distribution of harm. Slow violence disproportionately affects poor and racialized communities (Bullard 1983). It is often intentionally practiced by those who wield power on those who are least able to contest and resist through institutional channels of accountability. Perpetrators of slow violence can get away with inflicting harms on people and places that have little recourse to respond, thereby further marginalizing these communities and landscapes. Harm to landscapes incorporates an additional dimension to this kind of violence, which is perpetuated by humans on nonhuman species.

The differential experience of communities and landscapes subject to slow violence is often highlighted in the literature on 'zones of sacrifice.' This was a label first used by United States government officials during the Cold War to identify areas with significant levels of contamination due to nuclear testing and uranium processing. The term has been used more widely in environmental justice research to capture the 'irrefutable empirical evidence that not all American communities are created equal' (Bullard 2011, p. A266). Zones of sacrifice encompass communities and landscapes that have been designated by the state as expendable. In these areas, minority populations co-exist with heavily polluting industrial activities aided by the complicit silence of political authorities (Davies 2019). For example, Lerner's (2010) research in low-income, minority communities in the United States that live with disproportionately high levels of contamination and resulting health damages documents how this coexistence is protracted by government administrations that have regarded these areas from their start as the sacrifice that citizens needed to make at the altar of national security and prosperity.

Heterogeneity of marginalized populations

Although they are often characterized as a single class, marginalized populations are not a homogenous group. Broad measures of inequality based on GDP gloss over different dimensions of vulnerability. However, even specific categories, such as being elderly, also require us to consider that not all elderly people are vulnerable, nor are they vulnerable in similar ways. In the case of gender, there is evidence that women suffer disproportionate harms in mining communities including through domestic violence and prostitution, yet here again gender based violence can vary according to proximity to mining camps and economic opportunities (Lahiri-Dutt 2011).

There is also a geographic dimension to the nature and intensity of harms experienced through slow violence. For example, based on ethnographic research along the Chad-Cameroon oil pipeline, Murrey (2016) examined variations on experiences across communities that had ties to the ruling family versus those that did not. In addition, she found that 'community members and communities are affected unevenly, depending on their location along the pipeline, proximity to the pipeline within the village, gender, age, political connections or ethnic heritage' (242). The same environmental threat traverses diverse spaces, affecting households and communities differently because each is subject to pre-existing social, economic and political opportunities. Given this heterogeneity, research on vulnerability needs to focus on contingencies that are better addressed through case studies at lower, rather than higher, levels of analysis. Differential impacts on households and communities offers evidence of where and why vulnerability exists, allowing for more effectively tailored policy responses.

Visibility of violence

Although communities may exercise agency by making political demands of their government representatives, launching protests and staging refusals to delay harmful industry projects, as I explore in more detail in the next section, one of the key features of slow violence is its very limited visibility to outsiders. Nixon (2011) offers the example of domestic abuse and the experience of victims who suffer the protracted, often bloodless but no less devastating effects of continuous violence. Yet, who bears witness to this violence, and presents it publicly to mobilize society and accountability mechanisms in favor of its victims? Unlike media coverage of fast and explosive kinds of environmental disasters, Nixon argues that stories of bloddless and slow motion violence tend to be buried in the news 'particularly if it's relayed by people whose witnessing authority is culturally discounted' (Nixon 2011, p. 16). Therefore, slow violence remains mostly invisible (Rice 2016).

However, some scholars have argued that it is important to differentiate to whom these kinds of slow motion shocks are visible (Cahill and Pain 2019, Davies 2019). For example, Davies' 2019 ethnographic research on the toxic environments produced by petrochemical plants along the Mississippi River in the United States, in what is known as Cancer Alley, reveals that the onset of pollution is very visible to affected communities. The problem is not that this kind of violence is out of sight but that it is only visible to people and places that have already been discounted. Davies characterizes this as 'a politics of indifference about the suffering of marginalized groups [that] helps to sustain environmental injustice, allowing local claims of toxic harm to be silenced' (Davies 2019, p. 13). Frequently, women become leaders or spokespersons of communities affected by this type of violence and rise up to demand environmental and health protections (Lerner 2010, Nixon 2011). In Latin America, women environmental defenders have become a strong symbol of resistance to extractive industries (Muñoz and Del Carmen Villarreal 2019). However, their ability to achieve changes that mitigate significant risk to communities and landscapes is tempered by the degree of recognition and participation in the broader sociopolitical system in which they live, and their gender also mediates that recognition. In the next section, I discuss how ongoing stressors dispossess communities of their means to respond to fast and slow moving shocks.

Responding to environmental violence

The literature on environmental change has underexamined questions of agency in response to environmental shocks. Brown and Westaway argue that both scholarly and policy approaches have predominantly centered on the provision of infrastructure and interventions that are born outside affected communities and are 'based on simplistic assumptions between poverty and adaptive capacity' (2011, p. 321). What is most often missing is a disaggregation of individual and community capacities to activate resources in response to environmental disasters.

In the environmental justice literature, capacities to respond to environmental violence have been discussed in relation to distribution of burdens, recognition, influence in decision-making, and capabilities (Schlosberg 2004, Schlosberg and Carruthers 2010). Capabilities in this context are defined as the extent to which people have the means to realize lives of dignity, which they have reason to value and enable them to function and flourish (Sen 2009, Edwards *et al.* 2016). In addition to methodological challenges with operationalizing capabilities in quantitative studies, a further concern has been with the overreliance on individual wellbeing rather than collective or community capabilities where power differentials can produce vulnerability variations (Naz 2016).[1] Furthermore, capacities to respond are not only historically contingent but also subject to site specific patterns of social, political and economic development that circumscribe agency. These concerns have animated the literature of social vulnerability and community resilience (Norris *et al.* 2008).

Capacities to respond have also been conceptualized in terms of the enfranchisement individuals enjoy in a given sociopolitical and economic system, and their legitimate participation in decisions over entitlements (Sen 1982, Appadurai 1984, Bohle *et al.* 1994). In addition to participation, enfranchisement presupposes recognition of the groups whose interests need to be considered and denotes stakeholders' influence over agenda setting and decision-making with public or private authorities. A family living in the midst of an oil spill should not only participate in consultations regarding clean-up efforts, but first be recognized as a key stakeholder in the governance of oil projects around them, be able to influence decision-making on extractive projects, and finally participate in any remediation process. Hence, recognition, influence and participation are key mechanisms for enfranchisement. I refer to enfranchisement as a nuanced and variable position in society rather than a static and binary condition – and one that goes well beyond the right to vote. Enfranchisement has psychosocial attributes at the individual and community levels, which are discussed next.

Individual capacities

Some types of violence create a psychological footprint that limits opportunities for enfranchisement. Slow violence achieves disenfranchisement through exhaustion, fear and shame in addition to material displacements in the form of physical dispossession from land, environmental and human health degradation, and dependent development (Kramarz *et al.* 2021) . People who live in sacrifice zones highlight psychological dimensions. However, this experience is not limited to the poor. The emotional toll of slow violence limits the agency of people who may enjoy other kinds of capacities, including economic resources.

To illustrate, I refer to two communities living adjacent to oil sector activities. At one end of the spectrum, people in a shantytown named Flammable in Argentina live next to one of the most polluted watersheds in the world, subsisting in precarious conditions in terms of housing, employment and health access (Kramarz 2018). At the other end of the spectrum is a community in the Dallas – Fort Worth area of Texas that sits on top of the Barnett Shale – a rich natural gas field accessible through fracking. Gullion (2015), an epidemiologist by training, conducted an ethnography in affected areas of Texas to examine what she calls the reluctant activism of residents who are largely educated, middle class and politically conservative – and would never identify themselves as environmentalists.

Though the Texas residents are much more economically enfranchised than their counterparts in Flammable, what is common to both communities is the emotional impact of slow violence and how that experience limited or fatally delayed their capacities to respond to it. The residents in Texas provided Gullion an extensive account of their feelings of fear from all the illnesses they were experiencing, anger, distrust, sense of betrayal by government regulators, frustration and powerlessness. This slowed down their response to the fracking pads that were being built all around them. The psychological impact of being disbelieved, discredited, ignored by government officials and corporate representatives delayed the mobilization of these middle-class activists to organize and elect local city council members who would help restrict and regulate drilling.

Hirsch *et al.* (2018) found that fracking exerts various impacts on individuals' mental health. During the planning process, it can increase individuals' feelings of powerless, due to a sense of fear and betrayal as they anticipate the prospect of fracking entering their community (Short and Szolucha 2019). The disruption caused by fracking operations creates higher amounts of psychological stress and anxiety(McDermott-Levy and Garcia 2016, Watterson and Dinan 2018). Fracking can be associated with physical illnesses and neurological problems, which further cause adverse effects on mental health due to the feeling that both this 'fracking-induced illness' and its remedy are out of one's control or reach (Chapman *et al.* 2004). Scholars have also documented how individuals' sense of powerlessness can contribute to a dissociation from political participation and advocacy, reducing chances for mobilization in sacrifice zones (Dunlop, Atkinson, and Diepen 2020). In 'Capitalism and the Politics of Resignation,' Benson and Kirsch (2010) discuss the phenomenon of resignation following the Ok Tedi mining disaster in Papua New Guinea, where people living downstream submitted themselves to their fate after learning that the damages along the river would last from 60 to 100 years.

For the residents of Flammable in Argentina, fewer economic resources meant that, in addition to the emotional toll of being ignored by public officials, families had to deal with the cognitive load of poverty as they attempted to mitigate the environmental destruction that surrounded them. Yet, what kind of response to environmental disasters can a family generate when it is living on a lot without a property title, lacking municipal connections to water and sanitation, and attending to their children and parents' own chronic sicknesses from environmental contaminants? Psychologists who study constraints on mental bandwidth under conditions of scarcity argue that cognitive capacity changes with different economic circumstances (Schilbach *et al.* 2016). For example, being poor means that more thinking is devoted to finding ways to meet immediate needs with limited or sporadic income. This distracts individual attention that might otherwise be given to lobbying political representatives or gaining information about sources of contamination. In other words, 'when you are poor, economic challenges are more than just economic, they are also cognitive' (Schilbach *et al.* 2016, p. 438).

Community capacities

In addition to individual capacities, communities provide the foundation for networked resources. Norris *et al.* (2008), incorporating a public health perspective, focus on the concept of wellness as their preferred outcome variable. This is a rejection of resilience as an end in itself. They develop a framework that consists of four sets of resources that shape community responses to ongoing stresses and acute crisis. These are economic development, social capital, information, and community competence. For Adger (2000), economic development refers to growth, stability and distribution of income. Stability is especially important in the case of resource dependent communities because commodity markets are subject to wide pricing variations that increases risk. This is particularly the case in the absence of diversified economies, like many in the Global South that are largely dependent on a few extractive industries (Kingsbury *et al.* 2019).

Access to political institutions and representatives is another key variable. For instance, in the distribution of post disaster resources, Norris *et al.* (2008, p. 137) claim that allocations abide by need but often 'support follows the "rule of relative advantage" because one's embeddedness in the community, political connections, and social class determine the availability and accessibility of resources.' For example, the response to Hurricane Andrew in South Florida, albeit not an extractive industry disaster, offers a glimpse of an event whose recovery was documented as 'inherently competitive' and based on relative advantages along a socio political scale (Peacock *et al.* 1997, p. 17).

Social capital encompasses all those resources that are connected to the development and maintenance of durable social networks that facilitate daily life and mutual aid in situations of need. In the case of the Argentine community of Flammable, described above, dependable relationships between neighbours encouraged many people to stay even as contamination from the nearby petrochemical industries worsened environmental and health conditions. An ethnographic study of the community describes the process of deciding to stay as 'crucial to understand how they think and feel about this (contaminated) place – not in the way an outsider would but in a way that is thoroughly embedded in history and the routine organization of daily life' (Auyero and Swistun 2009, p. 86). Social support is also engendered through 'community bonds, roots and commitments' (Norris *et al.* 2008, p. 139). While deep attachments to place can enhance capacities for community responses, disasters that dislocate people from their communities can be particularly destructive to those same capacities. According to Norris *et al.* (2008), information and communication are also key components of effective community responses and this includes the ability to construct a shared narrative for social mobilization. Whereas the last set of resources are grouped under the comprehesive label of 'community competence,' Norris *et al.* (2008) refer more specifically to capacities for diverse actors to organize and advocate for their shared needs.

The structure of vulnerabilities: shocks and capacities to respond

The relationship between exogenous shocks and endogenous responses does not follow a linear, causal pathway. It is a relationship of mutual constitution. Different types of vulnerability profiles emerge when different kinds of violence interact with capacities to respond. Furthermore, since shocks have an unequal distribution of impacts, communities that start with limited means to respond tend to be left worse off with each additional shock. When communities have limited means to respond or hold authorities accountable, the kind of remedies that follow environmental disasters tend to be marginal adaptations rather than transformational shifts in risk and exposure (Ribot 2011).

In this section, I connect the previous analysis on fast and slow violence with a discussion on capacities to respond, to examine the kinds of vulnerability profiles that emerge with different types of structure and agency conditions. What follows is a typology that describes the creation of different kinds of vulnerable subjects. Each scenario is characterized by varying degrees of visibility and opportunities for more or less effective responses from communities, states, industries and civil society organizations beyond the affected populations. Each scenario discussed below is named for the kind of vulnerable subject it creates. The scenarios illustrated in Figure 1 are dynamic since vulnerability profiles can shift with each new disaster. For example, a slow and steady diet of harms visited on a community that has lived next to a pipeline but remained largely outside the public eye (scenario 1) might lead to a massive oil spill that receives widespread media attention and becomes temporarily visible (scenario 3). Mobilization on behalf of affected communities by outside actors who enjoy higher levels of enfranchisement can create broader political constituencies that shift the opportunities to respond to the disaster again (scenario 4). This movement within the matrix is not teleological in the sense of progressing towards a world of less exposure to violence or greater stakeholder capacity to respond to its harms. The catastrophic rupture of a mine tailings dam might give more visibility and amplify a community's calls for political action but the event might recede from the public consciousness soon thereafter, tapering off mobilizations and returning the community to an even worse-off version of its pre-disaster status – one that continues to be shaped by slow violence on top of a catastrophically devastating event. A double arrow along both axes of the matrix represents the non-static and non-teleological quality of the typology below.

Scenario 1: discounted and invisible subjects

The first configuration of vulnerability is constituted by the ongoing risks of slow violence and low levels of enfranchisement among the affected stakeholders. There is contestation over the harms generated by this violence, and political claims are presented to governments, corporations, and

		Capacity to Respond	
		Low	High
Environmental Shock	Slow	Discounted and Invisible Subjects ①	Subjects of Doubt and Stretched Assets ②
	Fast	Subjects of the Environmental Moment ③	Subjects of Significant Risk Reversals ④

Figure 1. Subjects of vulnerability: Environmental shocks and capacities to respond.

non-state actors but the political invisibility and economic marginalization of the affected communities weakens the influence of these demands in generating changes that can lower their risk exposure. For example, in the case of the communities living along the Chad-Cameroon oil pipeline, Murrey's (2016) informants said that the company's community relations officer would always see them to hear their complaints about the construction project but they would not receive any response to those complaints or witness any changes to avert the contamination of their water supply, or destruction of their plantations. Murrey concludes 'not only were people unsure about whom to approach, they were doubtful that complaints would be seriously considered, because of the state's endorsement of the project' (243). Yet, Murrey's work also highlights that vulnerable communities are not without tools of resistance, but by necessity they employ the weapons of the structurally weak. After prolonged experiences with slow violence, which in her research included land appropriation, water contamination, lack of consultation, and de facto absence of any effective complaint mechanisms, each community developed its own strategies relying on what Murrey calls 'slow dissent.' For example, some villages start by refusing to receive pipeline company representatives, and some households refused to help the company contract laborers to work on the pipeline's right of way. These refusals amount to slow, dispersed, often unorchestrated acts of dissent (Scott 1985). This dissent can often remain symbolic – by denying the granting of authority and legitimacy to decision makers.

In their ethnography of Flammable, Auyero and Switsun (2007, 2009) also describe the fear that residents continuously felt about being relocated and rumors about what Shell, the largest company in the compound, might do next. The authors conclude that residents 'are playthings of environmental, economic and political misfortunes – hardly of their own making' (2007, p. 126). Their research documents the devastating effects of toxic contamination on bodies and minds, where environmental damage has been normalized as landscapes have been implicitly or explicitly designated as sacrifice zones.

This scenario creates discounted and invisible subjects. Government or corporate responses remain largely symbolic, such as participation through consultations in ways that provide cover and delay effective protections. The International Finance Corporation and Exxon Mobil, which are part of the Chad-Cameroon pipeline consortium, publicised an extensive level of consultation, calling it unprecedented and possibly the largest undertaken in the world for such a project (Murrey 2016). In reality this did not translate into empowerment. This is the most vulnerable of all scenarios because exposure aggregates over time, the impacts are registered after a critical number of people and ecosystems have been allowed to die, and the trajectory to that tipping point emotionally and physically devastates communities and landscapes that are already the most marginalized within the existing economic and political system.

Scenario 2: subjects of doubt and stretched assets

The second configuration shares the characteristics of slow violence with the previous scenario. There is a similar emotional and physical toll on people in these communities who manifest fear and confusion; fear about their sicknesses and uncertainty about the claims that government officials and industry representatives present to them. This doubt can splinter communities that further undermines effective responses to the environmental disaster. For example, in Centralia, Pennsylvania a mine fire in a coal seam that started in 1962 and continues to burn underground to this day generated 'ambiguous evidence and contradictory signals' that ended up eroding any measure of social capital the community could have leveraged to mobilize effective redress for health harms families were suffering (Kroll-Smith and Couch 2009). Stretched beyond their capacities to advocate as a community, everyone in Centralia was eventually relocated to different places thereby terminating the existence of the community altogether.

In other cases, and after delays brought on by doubt, communities become internally organized. They might have some epistemic and economic capital of their own, as described in Gullion's (2015) account of the communities in Texas living next to fracking pads. Or they might be assisted by outside organizations (domestic or international) that they rely on for legal advice, public communication and other strategies of resistance. This collaboration between domestic and international activists can trigger a boomerang pattern of influence on state behavior (Keck and Sikkink 2014). This makes communities more visible to journalists and non-governmental organization in their domestic setting. More importantly, efforts at collective action in partnership with transnational activists makes communities and landscapes visible to an international audience. This is a critical target for bearing witness to violence in states where democratic institutions are not consolidated, and accountability mechanisms do not work in practice. Access to international organizations like the Inter-American Court of Human Rights, or non-judicial bodies like complaint mechanisms of multilateral development banks that were involved in projects causing harms, are also among the repertoire of possible responses in this scenario.

Scenario 3: subjects of the environmental moment

The third scenario emerges from the kind of violence that, by its nature as 'immediate in time, spectacular and explosive in space' (Nixon 2011, p. 2), becomes instantly visible and difficult to hide. However, here too the level of enfranchisement of affected stakeholders will often render that visibility temporary. When the affected stakeholders or landscapes are not deeply entrenched into the political and economic system of the state and market, they lack ongoing recognition so that visibility and resources offered to the community are short-lived. Given the lower economic development and political enfranchisement that define these communities, they are particularly ill prepared and very dependent on outside resources to respond.

Due to the sensational nature of the violence, which features extensive demands yet transitory responses, this scenario creates subjects of the environmental moment. These cases may incite mass mobilizations on social media or in the streets. They also frequently enjoy the attention of scholars. Whereas the vastness of the violence makes it compelling to different audiences, the long-term prospects for effective responses among those affected are not significantly altered by the spotlight. An example is the Mariana mining disaster of 2015 in Brazil – the largest environmental disaster in that state and one where reparations were eventually left in the hands of industry, rather than governments, to determine who was affected and at what cost. Reparation decisions where communities are sidelined undermine social cohesion. For example, the BP Deepwater Horizon oil spill in 2010 in the Gulf of Mexico caused immense habitat destruction, and loss of income to fishing and tourism industries. Many communities also experienced what psychologists have called 'community corrosion' due in part to different amounts of damage settlements from BP to individual households (Mayer *et al.* 2015).

Scenario 4: subjects of significant risk reversals

The fourth scenario is one of least vulnerability. That does not mean that communities in this scenario are shielded from massive violence, but that this configuration of vulnerability has a comparative advantage over others discussed above. Affected communities enjoy greater levels of enfranchisement and this privileged standing in a political and economic system they can access enables them to make governments and corporations more responsive to their demands. Subjects with opportunities to redress the sources of a visibly catastrophic event enjoy the benefits of visibility and access to economic, political and social capital that assists them in generating community responses and marshalling the help of outsiders. It is a scenario that has the highest potential for generating significant risk reductions and transformational changes beyond the affected communities through legislation, accountability procedures or financial reparations that can mitigate the likelihood of future disasters. A feature of this scenario is that it is the most likely to generate responses that are not just reactive to the disaster itself but also proactive to minimize future exposure (Gallopín 2006, Kasperson *et al.* 2012).

Hernan (2010, p. 7) argues that it often takes a fast onset environmental disaster to achieve this shift and 'overcome the stagnation and vested interests that block regulatory reform.' Some of the examples he offers are the community mobilizations that followed the London Fog of 1952, which was caused by coal dust and led to the British Clean Air Act in 1956. Another is the 1976 explosion in a chemical plant owned by the Roche Group, which released a toxic chemical cloud of dioxin poisoning the environment and human populations in the Lombardy region of Italy, and concentrated around the town of Seveso. This led to the European Union's adoption of the Seveso Directive that regulates chemical and petrochemical industries. Beyond these legal ramifications, environmental disasters like these have the potential to become ideational turning points. In the case of Seveso, the disaster emerged as a community symbol of broader environmental struggles, and has shaped local identities that mobilize in support of green policies evoking the legacy of the disaster (Centemeri 2010).

Conclusion

The concept of vulnerability has been deployed in ways that tend to 'flatten and simplify diverse communities' (Marino and Faas 2020, p. 33). I have argued for a more nuanced conceptualization of social vulnerability, one that is contingent on the nature of the shocks and the agency of communities to respond. Not all shocks are similar. In the case of slow violence, communities that may otherwise be considered economically enfranchised are, on closer analysis, dispossessed of capacities to respond because of the psychological toll they experience through slow violence. In the case of economically disenfranchised communities, shocks can become devastating because those affected were already functioning at the edge of their capacities to survive. Structure and agency intersect in the creation of vulnerable subjects. It is important to understand these processes and conditions that underpin different types of vulnerabilities because the focus of disaster prevention policies is to individualize resilience and direct responsibility away from structural and systemic forces that create vulnerable subjects. This misdirects the onus of responsibility and locus for action.

I offer a typology in this article to help develop insights into the underlying dimensions of social vulnerability and make it possible to better classify cases of vulnerable populations (Collier *et al.* 2012). Typologies help to create analytical categories that provide a starting point for thick, qualitative research to understand the interplay of factors that shape community experiences on a case by case basis. Methodologically, this cannot be gleamed from quantitative cross-country comparisons that do not incorporate subjective community voices, values and histories. There have been some incipient calls for slow research, a term borrowed from the slow food movement that highlights local experience as a starting point for analysis (Heitmann *et al.* 2011, Adams *et al.* 2014). 'Like the pause before the meal, ethnographic pauses during which the researcher listens, engages, and observes patiently, enable us to understand the process, history, and genesis of the event' (Adams *et al.* 2014, p. 187). This recursive approach to our scholarship is likely to yield transformative insights to researchers and subjects of study alike.

Note

1. There is considerable discussion in the environmental justice literature regarding how capabilities ought to be conceptualized in relation to individual versus social goals. Liberal approaches privilege individual well-being while radical approaches focus on community well-being and collective goods (see Edwards *et al.* 2021).

Disclosure statement

No potential conflict of interest was reported by the author(s).

References

Adams, V., Burke, N.J., and Whitmarsh, I., 2014. Slow research: thoughts for a movement in global health. *Medical Anthropology*, 33 (3), 179–197. doi:10.1080/01459740.2013.858335

Adger, W.N., 1999. Social Vulnerability to Climate Change and Extremes in Coastal Vietnam. *World Development*, 27 (2), 249–269. doi:10.1016/S0305-750X(98)00136-3

Adger, W.N., 2000. Social and ecological resilience: are they related? *Progress in Human Geography*, 24 (3), 347–364. doi:10.1191/030913200701540465

Adger, W.N. and Mick Kelly, P., 1999. Social vulnerability to climate change and the architecture of entitlements. *Mitigation and Adaptation Strategies for Global Change*, 4 (3), 253–266. doi:10.1023/A:1009601904210

Appadurai, A., 1984. How moral is South Asia's economy?— a review article. *The Journal of Asian Studies*, 43 (3), 481–497. doi:10.2307/2055760

Auyero, J. and Swistun, D., 2007. Confused because exposed: Towards an ethnography of environmental suffering. *Ethnography*, *8* (2), 123–144. doi:10.1177/1466138107078630

Auyero, J. and Swistun, D.A., 2009. *Flammable: environmental suffering in an argentine shantytown*. Oxford: Oxford University Press.

Barrett, G., 2015. Deconstructing Community. *Sociologia Ruralis*, 55 (2), 182–204. doi:10.1111/soru.12057

Benson, P. and Kirsch, S., 2010. Capitalism and the politics of resignation. *Current Anthropology*, 51 (4), 459–486. doi:10.1086/653091

Blaikie, P., *et al.*, 1994. *At risk: natural Hazards, people's vulnerability and disasters*. London: Routledge.

Bohland, J., Davoudi, S., and Lawrence, J.L., 2018. *The resilience machine*. New York: Routledge.

Bohle, H.-G., 2001. Vulnerability and Criticality: perspectives from Social Geography. *IHDP Update*, 2 (1), 3–5.

Bohle, H.G., Downing, T.E., and Watts, M.J., 1994. Climate change and social vulnerability: toward a sociology and geography of food insecurity. *Global Environmental Change*, 4 (1), 37–48. doi:10.1016/0959-3780(94)90020-5

Branch, A., 2018. From disaster to devastation: drought as war in Northern Uganda. *Disasters*, 42 (S2), S306–27. doi:10.1111/disa.12303

Brown, K. and Westaway, E., 2011. Agency, capacity, and resilience to environmental change: lessons from human development, well-being, and disasters. *Annual Review of Environment and Resources*, 36 (1), 321–342. doi:10.1146/annurev-environ-052610-092905

Bullard, R.D., 1983. Solid waste sites and the Black Houston community*. *Sociological Inquiry*, 53 (2–3), 273–288. doi:10.1111/j.1475-682X.1983.tb00037.x

Bullard, R.D., 2011. Sacrifice zones: the front lines of toxic chemical exposure in the United States. *Environmental Health Perspectives*, 119 (6), A266.

Cahill, C. and Pain, R., 2019. Representing slow violence and resistance. *ACME: An International Journal for Critical Geographies*, 18 (5), 1054–1065.

Centemeri, L., 2010. The Seveso disaster legacy. *In*: *Nature and history in modern Italy*. Athens, OH/M. Armiero and M. Hall: Ohio University Press & Swallow Press, 251–273. Available from: https://hal.archives-ouvertes.fr/hal-01016045.

Chambers, R., 1989. Editorial introduction: vulnerability, coping and policy. *IDS Bulletin*, 20 (2), 1–7. doi:10.1111/j.1759-5436.1989.mp20002001.x

Chandler, D., 2019. Resilience and the end(s) of the politics of adaptation. *Resilience*, 7 (3), 304–313. doi:10.1080/21693293.2019.1605660

Chapman, D.P., Perry, G.S., and Strine, T.W., 2004. The vital link between chronic disease and depressive disorders. *Preventing Chronic Disease*, 2 (1), A14.

Coburn, A.W., Spence, R.J.S., and Pomonis, A., 1991. *Disaster Mitigation*. UNDP Disaster Management Training Programme.

Collier, D., LaPorte, J., and Seawright, J., 2012. Putting typologies to work: concept formation, measurement, and analytic rigor. *Political Research Quarterly*, 65 (1), 217–232. doi:10.1177/1065912912437162

Collins, T.W., 2009. The production of unequal risk in hazardscapes: an explanatory frame applied to disaster at the US–Mexico Border. *Geoforum, Themed Issue: The 'View from Nowhere'? Spatial Politics and Cultural Significance of High-resolution Satellite Imagery*, 40 (4), 589–601. doi:10.1016/j.geoforum.2009.04.009

Cutter, S.L., 1996. Vulnerability to environmental hazards. *Progress in Human Geography*, 20 (4), 529–539. doi:10.1177/030913259602000407

Cutter, S.L., Boruff, B.J., and Lynn Shirley, W., 2003. Social Vulnerability to Environmental Hazards*. *Social Science Quarterly*, 84 (2), 242–261. doi:10.1111/1540-6237.8402002

Davies, T., 2019. Slow violence and toxic geographies: 'out of sight' to whom? *Environment and Planning C: Politics and Space*. April, 2399654419841063. doi:10.1177/2399654419841063

Dilley, M. and Boudreau, T.E., 2001. Coming to terms with vulnerability: a critique of the food security definition. *Food Policy*, 26 (3), 229–247. doi:10.1016/S0306-9192(00)00046-4

Dunlop, L., Atkinson, L., and Turkenburg-van Diepen, M., 2020. Perspectives on fracking from the sacrifice zone: young people's knowledge, beliefs and attitudes. *Chemistry Education Research and Practice*, 21 (3), 714–729. doi:10.1039/D0RP00022A

Edwards, G.A.S., Reid, L., and Hunter, C., 2016. Environmental justice, capabilities, and the theorization of well-being. *Progress in Human Geography*, 40 (6), 754–769. doi:10.1177/0309132515620850

Gallopín, G.C., 2006. Linkages between vulnerability, resilience, and adaptive capacity. *Global Environmental Change, Resilience, Vulnerability, and Adaptation: A Cross-Cutting Theme of the International Human Dimensions Programme on Global Environmental Change*, 16 (3), 293–303. doi:10.1016/j.gloenvcha.2006.02.004

Galtung, J., 1969. Violence, peace, and peace research. *Journal of Peace Research*, 6 (3), 167–191. doi:10.1177/002234336900600301

Grothmann, T. and Patt, A., 2005. Adaptive capacity and human cognition: the process of individual adaptation to climate change. *Global Environmental Change*, 15 (3), 199–213. doi:10.1016/j.gloenvcha.2005.01.002

Gullion, J.S., 2015. *Fracking the neighborhood: reluctant activists and natural gas drilling*. Cambridge, MA: MIT Press.

Heitmann, S., Robinson, P., and Povey, G., 2011. Slow food, slow cities and slow tourism. *In:* P. Robinson, S. Heitmann, and P.U.C. Dieke, eds., *Research Themes for Tourism*, London: Routledge, 114–127.

Hernan, R.E., 2010. *This borrowed earth: lessons from the fifteen worst environmental disasters around the world*. New York: Palgrave Macmillan.

Hewitt, K., 2019. *Interpretations of calamity: from the viewpoint of human ecology*. London: Routledge.

Hirsch, J.K., *et al.*, 2018. Psychosocial impact of fracking: a review of the literature on the mental health consequences of hydraulic fracturing. *International Journal of Mental Health and Addiction*, 16 (1), 1–15. doi:10.1007/s11469-017-9792-5

Kasperson, J.X., *et al.*, 2012. Vulnerability to global environmental change. *The Social Contours of Risk: Volume II: Risk Analysis, Corporations and the Globalization of Risk*. January, 245–285. doi:10.4324/9781849772549

Keck, M.E. and Sikkink, K., 2014. *Activists beyond borders: advocacy networks in international politics*. Ithaca, NY: Cornell University Press.

Kingsbury, D.V., Kramarz, T., and Jacques, K., 2019. Populism or petrostate?: the afterlives of Ecuador's Yasuní-ITT initiative. *Society & Natural Resources*, 32 (5), 530–547. doi:10.1080/08941920.2018.1530817

Kramarz, T., 2018. Using the courts to protect the environment in Argentina: accountability pitfalls when judges have the last word. *Case Studies in the Environment*, 2 (1), 1–13. doi:10.1525/cse.2017.000992

Kramarz, T., Park, S., and Johnson, C., 2021. Governing the dark side of renewable energy: a typology of global displacements. *Energy Research & Social Science*, 74 (April), 101902. doi:10.1016/j.erss.2020.101902

Krause, K., 2009. Beyond definition: violence in a global perspective. *Global Crime*, 10 (4), 337–355. doi:10.1080/17440570903248270

Kroll-Smith, J.S. and Couch, S.R., 2009. *The real disaster is above ground: a mine fire and social conflict*. Lexington, KY: University Press of Kentucky.

Kuhlicke, C., *et al.*, 2011. Contextualizing social vulnerability: findings from case studies across Europe. *Natural Hazards*, 58 (2), 789–810. doi:10.1007/s11069-011-9751-6

Lahiri-Dutt, K., ed. 2011. Gendering the field: towards sustainable livelihoods for mining communities. *Asia-Pacific Environment Monographs, The Australian National University*, 6 (1), 248.

Lerner, S., 2010. *Sacrifice Zones: The Front Lines of Toxic Chemical Exposure in the United States*. Cambridge, MA: MIT Press.

Marino, E.K. and Faas, A.J., 2020. Is vulnerability an outdated concept? After subjects and spaces. *Annals of Anthropological Practice*, 44 (1), 33–46. doi:10.1111/napa.12132

Mayer, B., Running, K., and Bergstrand, K., 2015. Compensation and community corrosion: perceived inequalities, social comparisons, and competition following the deepwater Horizon oil spill. *Sociological Forum*, 30 (2), 369–390. doi:10.1111/socf.12167

McDermott-Levy, R. and Garcia, V., 2016. Health concerns of Northeastern Pennsylvania residents living in an unconventional oil and gas development county. *Public Health Nursing*, 33 (6), 502–510. doi:10.1111/phn.12265

Muñoz, E.E. and Del Carmen Villarreal, M., 2019. Women's struggles against extractivism in Latin America and the Caribbean. *Contexto Internacional*, 41 (2), 303–325. doi:10.1590/s0102-8529.2019410200004

Murrey, A., 2016. Slow dissent and the emotional geographies of resistance. *Singapore Journal of Tropical Geography*, 37 (2), 224–248. doi:10.1111/sjtg.12147

Navas, G., Mingorria, S., and Aguilar-González, B., 2018. Violence in environmental conflicts: the need for a multidimensional approach. *Sustainability Science*, 13 (3), 649–660. doi:10.1007/s11625-018-0551-8

Naz, F., 2016. Understanding human well-being: how could sen's capability approach contribute? *Forum for Social Economics*. August, 1–16. doi:10.1080/07360932.2016.1222947

Nevins, J., 2003. Restitution over coffee: truth, reconciliation, and environmental violence in East timor. *Political Geography, Forum: Political Geography in Question*, 22 (6), 677–701. doi:10.1016/S0962-6298(03)00052-0

Nixon, R. 2011. *Slow violence and the environmentalism of the poor*. Harvard University Press. Available from: www.jstor.org/stable/j.ctt2jbsgw.

Norris, F.H., *et al.*, 2008. Community resilience as a metaphor, theory, set of capacities, and strategy for disaster readiness. *American Journal of Community Psychology*, 41 (1–2), 127–150. doi:10.1007/s10464-007-9156-6

Peacock, W.G., Morrow, B.H., and Gladwin, H., 1997. *Hurricane Andrew: ethnicity, gender, and the sociology of disasters*. London: Psychology Press.

Peek, L. and Stough, L.M., 2010. Children with disabilities in the context of disaster: a social vulnerability perspective. *Child Development*, 81 (4), 1260–1270. doi:10.1111/j.1467-8624.2010.01466.x

Peluso, N.L. and Watts, M., 2001. *Violent environments*. Ithaca, NY: Cornell University Press.

Popke, J., Curtis, S., and Gamble, D.W., 2016. A social justice framing of climate change discourse and policy: adaptation, resilience and vulnerability in a Jamaican agricultural landscape. *Geoforum*, 73 (July), 70–80. doi:10.1016/j.geoforum.2014.11.003

Räsänen, A., Lein, H., Bird, D., and Setten, G., 2020. Conceptualizing community in disaster risk management. *International Journal of Disaster Risk Reduction*, 45, 101485. doi:10.1016/j.ijdrr.2020.101485

Ribot, J., 2011. Vulnerability before adaptation: toward transformative climate action. *Global Environmental Change*, 21 (4), 1160–1162. doi:10.1016/j.gloenvcha.2011.07.008

Ribot, J., 2013. Vulnerability does not just fall from the sky: toward multi-scale pro-poor climate policy. *In*: M.R. Redclift and M. Grasso, eds., *Handbook on climate change and human security*. Cheltenham: Edward Elgar Publishing, 164–199.

Ribot, J., 2014. Cause and response: vulnerability and climate in the anthropocene. *The Journal of Peasant Studies*, 41 (5), 667–705. doi:10.1080/03066150.2014.894911

Rice, J., 2016. Slow violence and the challenges of environmental inequality. *Environmental Justice*, 9 (6), 176–180. doi:10.1089/env.2016.0019

Romero-Lankao, P., Gnatz, D.M., and Sperling, J.B., 2016. Examining urban inequality and vulnerability to enhance resilience: insights from Mumbai, India. *Climatic Change*, 139 (3–4), 351–365. doi:10.1007/s10584-016-1813-z

Schilbach, F., Schofield, H., and Mullainathan, S., 2016. The psychological lives of the poor. *American Economic Review*, 106 (5), 435–440. doi:10.1257/aer.p20161101

Schlosberg, D., 2004. Reconceiving environmental justice: global movements and political theories. *Environmental Politics*, 13 (3), 517–540. doi:10.1080/0964401042000229025

Schlosberg, D. and Carruthers, D., 2010. Indigenous struggles, environmental justice, and community capabilities. *Global Environmental Politics*, 10 (4), 12–35. doi:10.1162/GLEP_a_00029

Scott, J.C. 1985. *Weapons of the weak: everyday forms of peasant resistance.* Yale University Press. Available from: https://www.jstor.org/stable/j.ctt1nq836.

Sen, A., 1982. *Poverty and famines: an essay on entitlement and deprivation.* Oxford: OUP Oxford.

Sen, A.K., 2009. *The idea of justice.* Cambridge: Harvard University Press.

Shapiro, J. and McNeish, J.-A. 2021. *Our extractive age : expressions of violence and resistance.* Taylor & Francis. Available from: https://library.oapen.org/handle/20.500.12657/48472.

Short, D. and Szolucha, A., 2019. Fracking Lancashire: the planning process, social harm and collective Trauma. *Geoforum*, 98 (January), 264–276. doi:10.1016/j.geoforum.2017.03.001

Watterson, A. and Dinan, W., 2018. Public health and unconventional oil and gas extraction including fracking: global lessons from a scottish government review. *International Journal of Environmental Research and Public Health*, 15 (4), 675. doi:10.3390/ijerph15040675

Watts, M.J. and Bohle, H.G., 1993. Hunger, famine and the space of vulnerability. *GeoJournal*, 30 (2), 117–125. doi:10.1007/BF00808128

Ziervogel, G., Pelling, M., Cartwright, A., Chu, E., Deshpande, T., Harris, L., Hyams, K., Kaunda, J., Klaus, B., Michael, K., Pasquini, L., Pharoah, R., Rodina, L., Scott, D., and Zweig, P., 2017. *Inserting rights and justice into urban resilience: a focus on everyday risk. Environment and Urbanization,* 29 (1), 123–138. doi:10.1177/0956247816686905

Index

Note: Figures are indicated by *italics*. Endnotes are indicated by the page number followed by 'n' and the endnote number e.g., 20n1 refers to endnote 1 on page 20.